Asteroids And Nasa Dart Mission

First Ever Planetary Defense Test

(Crashing Of Nasa Dart Spacecraft Into An Asteroid)

William Hines

Published By **Zoe Lawson**

William Hines

All Rights Reserved

Asteroids And Nasa Dart Mission: First Ever Planetary Defense Test (Crashing Of Nasa Dart Spacecraft Into An Asteroid)

ISBN   978-1-77485-893-6

Legal & Disclaimer

The information contained in this ebook is not designed to replace or take the place of any form of medicine or professional medical advice. The information in this ebook has been provided for educational & entertainment purposes only.

The information contained in this book has been compiled from sources deemed reliable, and it is accurate to the best of the Author's knowledge; however, the Author cannot guarantee its accuracy and validity and cannot be held liable for any errors or omissions. Changes are periodically made to this book. You must consult your doctor or get professional medical advice before using any of the suggested remedies, techniques, or information in this book.

Upon using the information contained in this book, you agree to hold harmless the Author from and against any damages, costs, and expenses, including any legal fees potentially resulting from the application of any of the information provided by this guide. This disclaimer applies to any damages or injury caused by the use and application, whether directly or indirectly, of any advice or information presented, whether for breach of contract, tort, negligence, personal injury, criminal intent, or under any other cause of action.

You agree to accept all risks of using the information presented inside this book. You need to consult a professional medical practitioner in order to ensure you are both able and healthy enough to participate in this program.

TABLE OF CONTENTS

Chapter 1: Dart Mission

DART: Double Asteroid Redirection Test
A live show on the circle of diversion of space rocks is an important test NASA and various organizations want to carry out before any actual demand is identified. The DART mission of NASA is a demonstration of the latest innovations in motor impactors, that can cause a space rock alter its speed and direction. DART is the first space mission that shows the diversion of space rocks by using the impactor. The spacecraft launched on an SpaceX Hawk 9 rocket out of Vandenberg Space Power Base in California.

Send off:
November 23rd, 2021 at 10:00 p.m. PST
(November 24, 2021, 1:21 a.m. EST)

DART Effect:
September 26th, 2022 at 7.14 p.m. EDT

Didymos -- the Most Effective Ziel for DART's Main Goal
DART's goal is the space rock framework that is paired Didymos that signifies "twin" as in

Greek (and is the reason for "twofold" as the missions' title).

Didymos is the best candidate for the most memorable planetarium guarding attempt even though it's not in a position to crash into Earth and, therefore, poses an actual threat to our planet? The frame is made by two separate space rock: the larger rocks of space Didymos (width 780 meters or 0.48 miles), and the smaller moonlet rock Dimorphos (breadth 160 meters 525 feet) and is in the middle of the larger space rock.

The orbital period for Dimorphos in relation to Didymos will be 11 hours, 55 mins and the distance between the two focal points of the rock formations is 1.18 km (0.73 miles). The DART rocket is expected to impact Dimorphos nearly head-on, reducing the amount of time required for the moonlet of a space rock to orbit Didymos by a few seconds.

The Didymos framework appears to be an obscured double from Earth which implies that Dimorphos is in front of and behind Didymos when it circles the larger space rock from Earth. So, Earth-based telescopes are able to observe the typical variation in the beauty that is the joint Didymos framework in order to determine the Dimorphos circle.

After the effecthas been achieved, the same method will reveal the adjustments to circles of Dimarhos in relation to estimates prior to the influence. The plan for the DART influence for September 2022 was determined to be the time where Distance from Earth as well as Didymos is set to a minimum, in order so that it can be used to enhance the maximum ability to adapt perceptions. Didymos will, in all likelihood be around 11 million km (7 millions miles) away from Earth in the time when it is under DART's influence. DART influence, however telescopes around the globe are likely to join the global effort to assess the effect of the DART effect.

The DART show is meticulously created. Didymos's circle isn't meeting Earth's orbit anytime in the current forecasts and the thrust of energy DART transmits to Dimorphos is not too high and will not cause any harm to dimorphos' orbit. Its mass DART rocket as of the time when it comes into contact with Dimorphos is believed to be around 570 kilograms (1260 pounds) dependent on the amount of fuel used by the shuttle prior to the event of active effect.

Its weight Dimorphos hasn't been formally determined, but despite suspicions about the space rock's density and size and mass, the size and weight of Dimorphos is believed to be about five billion kg. Additional information about the Didymos space rock framework as well as DART's motor effect math that is arranged is available within this distribution.

In addition, the adjustment of the circle of Dimorphos by the motor effect of DART is designed to move its circle closer to Didymos. This DART mission is an example of the ability to solve the question of a possible influence from space risk, but is it an ideal idea to have one to be located?

DART is a spacecraft designed to alter the shape of an orbiting rock to test the limits of innovation. DART's goal is to make sure that the space rock doesn't pose dangerous to Earth. The space rock framework is a perfect test bed to determine if intentionally crashing an aircraft against a rock can be possible to alter its direction in the event that an Earth-breaking space rock be discovered in the future. Although no space rock greater than 140 meters has a great chance to come into contact with Earth within the next 100 years, approximately 40% of these space rocks were discovered at the time of writing in October 2021.

DART Accomplices and Partners
It is believed that the DART project is developed and managed for NASA via NASA's Johns Hopkins College Applied Physical science Lab. NASA's Planetary Guard Coordination Office is the primary authority for planetary safeguard exercises and helping to support this DART mission. Present U.S. accomplice foundations on DART include NASA Goddard Space Flight Center, NASA Johnson Space Center, NASA Langley Exploration Center, NASA Glenn Exploration Center, NASA Marshall Space Flight Center, NASA Kennedy Space Center the NASA Send off the Administrations Fly Drive Lab, SpaceX, Aerojet Rocketdyne, Lawrence Livermore Public Lab, Coppery College, Carnegie Science Las Campanas Observatory, College of Colorado, Las Cumbres Observatory, Lowell

Observatory, College of Maryland, New Mexico Tech with Magdalena Edge Observatory, Northern Arizona College, Planetary Science Organization as well as The U.S. Maritime Institute.

It is the DART Examination Group likewise incorporates foundations from all across and around the world.

Hera along with her and AIDA Global Cooperation

The Hera mission, which is a part of a program of the European Space Office's (ESA) space wellness and exercise for security, has been set to launch in 2024, and then rendezvous with the Didymos framework in 2026, four years after DART's impact.

In Hera's main goal, the main rocket as well as its two companion CubeSats will conduct definite research of the two space rocks with particular attention paid to the pit created by DART's crash and the exact confirmation that the rock is Dimorphos. Hera's post-impact examinations that are specific will greatly improve the planetary guarding information obtained by DART's Space Rock Avoidance Test.

Two missions, DART as well as Hera, are in the process of being designed and executed without restriction, and their combined efforts can help general information improve to an enormous extent. NASA's DART mission is totally focused on global participation and ESA's Hera team members are invited to join as full participants from the DART group to contribute Dash's planetary guard exams and to help Hera realize its primary objective entirely.

Each of DART as well as Hera colleagues are crucial to the global collaboration called AIDA -AIDA - Space rock Effect and Diversion Evaluation. AIDA is the worldwide cooperation between the planetary guard and space rock scientists which will integrate the data obtained from NASA's DART mission, which includes ASI's LICIACube as well as the ESA's Hera mission to provide the most accurate details from the principal display of a space rock technology for avoidance.

AIDA is the joint effort by three groups: DART, LICIACube, and Hera groups, along with other analysts from around the globe to remove the best data needed for planetary guard and

Group Science from important space missions. The AIDA collaboration is a proof that planetary security is a global endeavor and that scientists and engineers from all over the world strive to solve issues related to global protections for the planet through international efforts.

Chapter 2: What We Learned From Nasa's Dart Mission

The spacecraft blasted through the space rock in order to illustrate how the space rock beneath Earth might be diverted.

It was the Double Asteroid Redirection Test shuttle (also known as DART was hit a small space rock, demonstrating innovation that could protect Earth from space rocks for a long time to come.

DART connects to Dimorphos, the rock in space. Dimorphos.

A space rock rebel is speeding towards Earth and is causing waves to collapse that cause mass destruction and the destruction of everyone on the planet.

Mankind has one chance to save the world with the bold, selfless, and selfless tales of guiding rockets into space to destroy that space rock.

However, this is not the story of. The night of Monday, NASA revealed what real world would look like. The space rock was there but it was not destroying the Earth. Furthermore, there was an aircraft, which was based entirely on the latest technology. The myths surrounding the mission actually took place at the design and physical science laboratory located in Baltimore as well as Washington, D.C.

Furthermore, there was an incident. This incident was the final mission in DART. Double Asteroid Redirection Test, also known as DART the spacecraft which was launched in November. It then raced through space for an extended duration as it searched for its goal -- a tiny space rock called Dimorphos 7,500,000 miles away from Earth.

"Interestingly humans have demonstrated the capability to autonomously modify and target the environment around an ethereal object," Ralph Semmel, director of the Johns Hopkins College Applied Physical science Lab, told an announcement following the incident. The lab handled the mission of NASA.

A space rock hit by a quick shot can slam its circle. In order for a space rock to attempt to reach Earth it could be enough to transform an immediate hit into an extremely close miss.

In the final minutes of its mission the rocket returned an array of images of the rock in space Dimorphos in the process of moving closer to it at over 14,000 miles per hour. DART was able to spot Dimorphos about an hour earlier, and it was it was a tiny speck of light. Then, the pile of precious rubble got larger and bigger until the surface of the space rock splattered with rocks appeared on the screen. The mission's engineers were on their feet singing and cheering.

"Typically the loss of signals from the spacecraft is horrible," Dr. Semmel declared. "Yet in this particular case it was the best outcome." There was a second half-image but the data did not return to Earth. DART crashed in the rock.

"Goodness it was amazing wasn't it?" stated Nancy Chabot, a planetary research scientist at the lab that manages the mission on the NASA webcast.

For a long time politicians needed to be able to justify their initiatives to protect our planet from the impact of space rocks. But, things began to shift as stargazers have had the chance to discover every one of the massive space rocks that could cause huge destruction, like those that bound the dinosaurs many years back, claimed Thomas Statler, the program scientist on DART. DART mission.

Global effects occur rarely, if every 10-million years, or more. But, because that probability is no longer a possibility, scientists at NASA and elsewhere offer their consideration to less significant items found in space. These are definitely more distinct and, despite the fact that they won't trigger massive eradications, they do generate more energy than an atomic bomb.

The emergence of a spotlight on planet guards will be seen in the various initiatives that NASA and legislators have backed. One of these is one of them, the Vera Rubin

Observatory, another telescope located in Chile which is sponsored through the US and is able to efficiently examine the night sky, and track the location of many potentially dangerous space rocks. Another one is one called the NEO Assessor, a space-based telescope which NASA is trying to create. It is also expected to find many dangerous space rocks, including those that are difficult to discern from Earth.

If any of the space rocks happen to be in a collision course with Earth The DART mission suggests that the possibility of diverting them is a good possibility.

The designers of the mission, who were employed in scientists from the Johns Hopkins College Applied Material Science Lab and the result was that 7.14 p.m. Eastern time, signified the end the work. The rocket, operating independently during the four hours of its existence was effectively locked onto Dimorphos.

It's a lot bigger than DART's camera was able to see Dimorphos fascinatingly a bit longer than an hour prior to the event. Dimorphos circles a larger stone, Didymos as well. Up until that point, the less imposing space rock was obscuring the light of the larger object. The DART route framework changed its gaze towards the less imposing space rock.

For up to five minutes prior to the influence of mission regulators, they could have been able to mediate if something resulted in a disastrous outcome. But, they had to not make any changes.

In the last five minutes, the people who were in control rooms were watching alike to the people who were watching the flood of images of Dimorphos. After that, the mission was over. The initial examination showed that the rocket landed within approximately 50 yards of the target area.

"I definitely have a better feeling," said Elena Adams an engineer for mission frames. "Also doing something that is that is this impressive is awesome. Additionally, we are eager to complete the project." The accident was planned, but it was triggered in the time that those space rocks appeared to be close to Earth.

The telescopes allowed Earth to take a clear view. The majority of them directed towards Didymos and Dimorphos according to NASA as well as the missions leaders. They also had also the Hubble along with the James Webb space telescopes just as cameras on Lucy Another NASA rocket.

The LICIACube is a rocket that's approximately similar to a shoebox operated for the Italian Space Organization, followed DART to capture photographs of the effects and the pile of garbage. Its direction was changed to the left so that it wouldn't collision with that space rock.

This allowed telescopes on Earth to take a clear view. The majority of them directed towards Didymos and Dimorphos according to NASA as well as the missions leaders. Also, the Hubble as well as the James Webb space telescopes just as those on Lucy the other NASA rocket. The LICIACube which is similar to a shoebox that was operated with the Italian Space Organization, followed DART's instructions to snap pictures of the effects and the garbage tuft. The direction of the rocket was changed to the right so it didn't also hit that space rock.

"There's the majority of us who anticipate the positive effects so that we can apply our science and try this," said Cristina Thomas who is a professor of cosmology as well as planetary science in Northern Arizona College and lead of the perceptions team gathering in preparation for the missions. "It is going to be so amazing and exciting once in an event of a blue moon that we're throwing everything that we can into it."

Over the next several long time frames over the next several long time frames, Over the next several long periods. Thomas and different stargazers will scrutinize the data and photos to figure out the way in which DART dealt with Dimorphos. The primary measure will be how much the less opulent space rock, which has been going around Didymos as if it was clockwork, and lasted for 55 minutes ago, has accelerated. This will be a reflection of the amount of force the rocket has given the space rock and made it move closer to Didymos. The effect is expected to be in the range of 1% or about seven minutes.

The impact could have caused Dimorphos an even greater push when the space rock had been an exact pile of rubble. The crash of DART could have caused an enormous pit, sending an explosion of debris into space. This gushing fountain of debris could have behaved as the force of a rocket engine striking against the rocks of space.

Stargazers want to know whether there is a light up of the Didymos-Dimorphos structure due to light bounced off the debris tuft. Analyzing the different colors of light might reveal clues about the structure of Dimorphos.

The more precise review will occur in the future when Hera, a spacecraft currently being worked on through the European Space Office, shows up to study these two rocks in space, specifically the mark left by DART. Researchers estimate that there should have been a mound that is 30 up to 60 inches wide.

Space rock couple takes two years in order to circumvent the sun, and a portion of the circle is a crossroads with Earth. However, there is no possibility of either of the space rocks striking Earth at any pointin time, and the result has zero chance of smashing onto Earth's surface.

However, with an impressive demonstration of how space rocks can redirect "I think that the earthlings should be able to rest better," Dr. Adams declared. "Certainly I'll."

Carolyn Ernst, a researcher who worked on the mission's camera, claimed that Dimorphos was akin to other rock types that spacecrafts from Earth have been able to take an examination as of late particularly that it looked like a pile of rubble, not the solid stone.

Space rocks found in the planetary cluster are the rough remnants from a planet that will never ever get mixed. Before they were able to join together due to the gravitational force of Jupiter destroyed them and scattered the fragments.

They are flawless pieces of rock typically unaltered over the last 4.5 billion years, and offering evidence of what the planet group was like prior to the existence of planets.

NASA and other space agencies have launched a large number of rockets that have visited diverse space rocks. That includes Japan's Hayabusa2 and NASA's OSIRIS-REX rocketthat took pieces of carbon-rich near Earth space rocks for a return focus on the planet. Hayabusa2 has taken off its models in December of 2020. OSIRIS-REX continues to return to Earth with a case filled with dirt and rocks from Bennu space rock Bennu is scheduled to parachute to arrive in Utah on September. 24, 2023.

Hayabusa2 and OSIRIS-REX aren't yet complete. Both rockets are but in the process of working, and they will also explore other nearby Earth satellites.

NASA has a second spacecraft, Lucy, that will be able to fly by a series of Trojan space rocks that were discovered in the orbit of Jupiter. The other NASA missions, Mind, is to explore its name the Mind space rock, which is rich in metal that could be the center of a smoldering protoplanet. The departure of Mind was scheduled for the present year but was postponed until the following year due to special problems.

Following the conclusion of DART the DART effect will end, and there will in all event be additional photographs of Dimorphos. A small satellite, also known as CubeSat which is called LICIACube will return images from the rock in the DART effect. The satellite, launched by DART prior to its self-destructive jumping was developed with the Italian Space Office and followed DART towards Dimorphos and then in a somewhat unusual method to ensure that it doesn't also hit Dimorphos's space rock.

In the month of October 2024, the European Space Organization is wanting to launch Hera as a new development for Dash which will study the effects of the effects on Dimorphos. Hera will arrive sooner than the 2026 deadline.

When NASA's gigantic Artemis, I moon mission is finally sent off one of the 10 CubeSats that will be hopping into space is NEA Scout, which will make use of sun-based sails to reach a close Earth surface. The exact goal will be contingent on the time when the rocket launched.

China has all the hallmarks of tackling an undertaking similar to DART. The month of July was when Andrew Jones, a writer who writes about China's Chinese Space program posted information on Twitter about a program that was hosted by an Chinese rocket maker.

Chapter 3: How Will Nasa Be Able To Know If Dart Worked ?

This is how NASA is going to be aware, assuming DART functioned.

When DART struck Dimorphos the space rock, it was guaranteed to alter the direction of the space rock due to the most fundamental laws of energy protection and force. If the circle around Didymos does not change in any way after the impact the Newton's laws of motion will be void.

DART can't overrule Newton.

If Dimorphos is strong and DART cuts it just a tiny of a distance, the cut follows the fundamentals of the Physical science 101 issue -two things impacting and remaining together. Because DART is moving towards the direction that is opposite to Dimorphos and is a drain on some of space rocks powerful force, which will cause it to move closer to Didymos and then accelerate.

However, if Dimorphos appears more like a pile of debris that is held together through gravity. impact will create a massive pit, sending a mass of debris out into space. This rock fountain will look like the push of a rocket's motor against the rock in space. In all likelihood the circle of Dimorphos (whose title is Greek meaning "having 2) will be closer towards Didymos (that term can be translated as Greek meaning "twin").

In the same way, as telescopes on Earth confirmed the existence of Dimorphos in the ongoing darkness when it was passing before and behind Didymos The same diminution will reveal its new orbital time.

It is believed that the European Space Office is sending an additional rocket named Hera that will go to Didymos as well as Dimorphos to look at the progress in both, focusing on that pit created by DART. The launch is scheduled for October 2024 , and arrive earlier than anticipated in 2026.

When the rocks from space grow up in cameras' feeds, the mission's will be focusing on the fact that no more orders can be delivered from the rocket. All in or all out at this time.

What happens if DART fails to detect?

NASA officials are optimistic they believe DART will not miss. "I'm absolutely certain that we'll be able to hit the target on Mondaymorning," Lindley Johnson, NASA's planetary guard, stated at a press conference last week, "and it will be an accomplishment that is completed."

NASA has been successful in deliberately throwing rockets into celestial bodies before. For instance LCROSS, the Lunar Pit Perception and Detecting Satellite also known as LCROSS was thrown into a pit near the south pole of the moon in 2009 and proved the existence of water ice at the south pole. Additionally there was the Profound Effect struck the comet Tempel 1 out of 2005 and sucked up garbage that was more dusty and colder than it was normal.

DART has a tougher to complete, since its goal which is its space rock Dimorphos is small and is only 500 feet wide. It was visible as a tiny dab in radar images made through Arecibo's radio telescope. Arecibo radio telescope however it is not visible to any optical telescope in the world of Earth is a bit small. Its size is an amount greater than an estimation of the ballpark rather as opposed to an estimation. the public isn't aware of the shape of its sphere.

The divine motion is more complex than it is more complex than the LCROSS or Profound Effect estimations. Dimorphos is located in a circle around a larger half-sized space rock called Didymos and has the distance of 0.6 miles that separates the two. Dimorphos completes one circle around Didymos in a manner that is reminiscent of clockwork. It takes about 55 minutes later.

"It is extremely difficult to strike a small object in space, but we'll get it done," said Elena Adams the DART mission engineering team's framework engineer. The camera at DART's will not recognize Dimorphos's presence as a distinct Dab from Didymos until approximately an hour prior to the collision.

DART is self-driving, self-destructing rocket that is taking it to its destination with those who are involved in the mission's workstations on The Johns Hopkins College Applied Material Science Lab in Maryland typically on the lookout for.

"You're going very fast," Dr. Adams declared. "Also when you reach that point you're not able issue any orders. So your framework has to be precise in the way it is controlling your rocket."

Dr. Adams said the mission group had 21 courses of action in the event that different events happen to go wrong. "The main reason for having members of the mission in the first place is to allow them the ability to intervene in the event of a crisis," she said.

In the event that Dimorphos proves to be smaller and less bright than we had hoped the time to open the camera of the rocket could be extended.

"We have 21 possibilities,"" Dr. Adams said. "Number 21 is an unrealized result."

In the event that this happens, "We will plunk down once more in our seats, and then start saving all of the documents on the board that explain the reasons we didn't," Dr. Adams stated.

The rocket may fail to hit the space rock for many reasons. In the case of example, if the drive framework suffered an unforgiving experience, it will likely result in the failure of the mission. However, if the problem was caused by a large beam of wanderers which caused an untimely restart of the computer of DART and the rocket was safe, there may be options to try once or with a different goal.

There are other ways of stopping space rocks that are rebelling.

DART shots aren't necessarily the only way to redirect space rocks.

"Different concepts have been analyzed," Lindley Johnson, the official in charge of the planetary guard at NASA in the news conference just a few days ago. "We have a list about three or four concepts that may be a possibility later try to evaluate."

One of them is commonly referred to as gravity farm trucksthe name refers to a heavy rocket which is that is launched to be surrounded to an object in space. This causes an inverse gravitational pull to the space rock towards the spacecraft. If there is a long enough time it could be enough to lift the space rock away from the crash course of Earth.

"The common fascination between the rocket and space rock will eventually pull that space rock away from its direction of influence to an uninvolved one," Mr. Johnson stated. "A approach like that requires longer to carry out which is why we'd need to be aware of the option of carrying the process."

Another option is to use particle motors similar to the one that drove DART towards that Didymos Space rock.

Particle motors use electric fields to boost light emission iotas. They are typically the xenon. They generate just a bit of push, but they can discharge continuously for long periods which is similar to firing the Gatling weapon using the smallest of slugs that could be fired towards the rock. Similar to this gravity-driven truck method as a follow-up, the sluggish but regular modifications to the way the space rock moves could be feasible.

Another idea could be to draw the rock with paint and later let the sun finish the job. The color you choose can alter how the space rock moves in light and also the small strain caused by the photons from daylight bouncing off of the surface. This could also be enough energy to allow it to move for a long time.

The most popular option of Hollywood -- nuclear weapons -- may also be a possibility, but it could be a chaotic proportion after all other alternatives have been exhausted out, in case there was not enough time to pursue the slower methods.

Just blowing a space stone apart isn't the answer when each of the pieces ends up hitting Earth or, at the very least and releasing a similar amount of energy into the air , causing an explosion of shock wave.

Because space rocks aren't essential elements in terms of shape or form they are difficult to predict the effects of a bomb in detail. But, it's possible that it would be a possibility. Researchers last year reported the possibility that a secret space rock that was up to 330 feet in diameter could be destroyed by a single megaton nuclear device that could result in 99.9 per cent of the mass struck away from Earth's path in the event that the bomb explodes not less than two months prior to the impact.

What will we see once DART is in effect.

As soon as DART hits its space rock targets, we won't see anything. Instead, the flood of pictures of the small Dimorphos space rock getting more large in the event that the DART shuttle moves closer will be frozen.

If they win the designers will scream. There's nothing like an end of interchanges to confirm that it was a successful accident.

The final image that will require about two seconds prior to hitting, with the outer layer of space rock filling the camera's field of vision will be the last appearance ever created by Dimorphos on the night of Monday.

It's not the final image, but it will be the last.

The rocket that is limping through DART is a small spacecraft called The Light Italian CubeSat for Imaging of Space rocks, also known as the LICIACube. It was developed with the Italian Space Organization, the LICIACube was launched alongside DART during the first 9 months. after which it was separated and moved in a different direction , in an unsteady direction, which could miss Dimorphos.

The LICIACube will take pictures of DART's final stage and also the next pit. However, since it's so small -- and its receiving wire is small too -- it might be able to transmit information slowly via a weak radio transmission into the satellite radio dish that NASA's Profound Space Organization. It could be some time before the first images of the LICIACube are available.

About 40 telescopes on the globe and a few in space will be pointed towards Dimorphic and its space rock Didymos as the effect takes place. They will not be able to observe Dimorphos and, in particular, the divot DART will create.

But, the assumption is that the framework of Didymos-Dimorphos will improve within the time following the impact.

"What we're trying to find is a general illumination of the entire structure, which will show how much waste and other debris was discharged because that Ejecta is released into space, and is also illuminated with the light of the sun." stated Thomas Statler the DART program's lead researcher, during an announcement meeting that took place last week.

The amount and speed at which the light up process happens "is an indicator of something to do with the consistency of the material lifted as well as how much it was" Statler said. Statler said.

The telescopes are NASA's Hubble and James Webb space telescopes as well as the camera mounted on the Lucy rocket which is designed to make an attempt to meet with space rocks that are caught in the ring of Jupiter. The Lucy mission was launched around a month prior to Dash.

Hubble will not be able to see on Didymos in the hour of the event because Earth will be visible. In all likelihood it will start looking around 15 minutes following the event. "That is okay since we don't expect that anything will be able to be detected by the precise photo of the effect" Statler said. Statler said.

If you can use the Webb telescope which spends much of its energy in observing distant worlds that are that are billions of miles away is able to follow a fast space rock that is less than 7 million miles away from Earth it's not clear.

"Allow me to move on, but this isn't the way J.W.S.T. is supposed to accomplish," Nancy Chabot, an planetary study in Johns Hopkins College in Baltimore who serves as the coordinator in DART DART mission, told the news conference on Sept. 12. "This is a very difficult estimate they have to work with."

In any event this is a worthy endeavor she added.
"This is a unique opportunity in a unique moment to use each and every asset that we have and boost the value we can realize," Dr. Chabot said, "so they'll examine. We'll see what they come up with."

The most important estimation is the modification of the time required for Dimorphos to complete the round around Didymos. The head-on collision of DART will drain a part of Dimorphos' rakish force leading it to fall closer to Didymos. It is predicted to accelerate Dimorphos and reduce the duration of its orbit, which is now at 11 hours, 55 minutes. This will reduce the orbit by about 11 percent.

"We will be able to see that the space rock paired framework is quick," Dr. Statler stated. This estimation, based on radar and the occasional darkness that occurs when Dimorphos moves ahead or behind Didymos is likely to take time.

"I am not sure the possibility of having a precise estimate of the period change in just a few weeks," Dr. Statler stated. "Also I'd be astonished if that it took more than the course of three months." The majority of the space rocks that are within the planetary group circle in the primary belt between Mars and Jupiter are never close to Earth.

However, other people, as a result of billions of years of gravitational do-sido, are thrown around in different circles. Some have been thrown into Earth's circle and therefore could potentially hit Earth at some point in the future. If the stargazers spot a rock in a collision course with Earth it would be necessary to stop it.

So, it is the DART mission.

Stargazers estimate that there are 250,000 of these close-to-Earth space rocks, which are larger than 460 feet in size. These are huge enough that a collision with Earth will produce more energy than even the most powerful nukes –- everything required to wipe out a city, however, not enough to trigger an enormous mass extermination.

(The space rock that was blamed for the end of the dinosaurs was much larger and was about six miles wide. The Stargazers have located them and none of them pose any danger for Earth.)

The other is Bennu the space rock that is as large in size as Realm State Building that NASA's OSIRIS-REX spacecraft recently visited. There's no chance to predict Bennu striking Earth within the next 100 years. But, in 2135, it'll be close, within 125,000 miles, or perhaps within the vicinity or around a small portion of the distance between the Earth toward the moon.

The weaknesses in precisely the way they're located will result in the mathematical simulations following 2135 more uncertain, and a significant portion of possible directions could hit Earth. In the year 2000, NASA determined a 1-in-1,750 chance of an impact in the next 2300 years with the risk increasing in the 2100s.

NASA continues to search for many space rocks which are believed to be at least 460 feet vast and in circles that are far from Earth. The other U.S.- funded ground-based telescope located in Chile is the Vera Rubin Observatory, will provide scopes of the night sky and will be able to see most of the missing space rocks. NASA is also working on an orbiting telescope known as NEO Assessor, which will also discover a variety of close Earth objects which include space rocks. At present, NASA is going for the gold by 2028.

In any case, when the 460-foot-wide space rocks are discovered however, the task isn't over. Impacts of smaller rock types, that are larger in size, can also cause serious harm. The meteor that struck Chelyabinsk, Russia, in 2013 was about 60 feet wide. It did not cause any passings but it left a number of wounds, mostly caused by glass that was broken through the impact wave.

NASA has recently started streaming pictures from the camera mounted on the DART rocket. It's integrated into the player for video above. At present only what you be able to see is a tiny dab, however, watch for that dab to grow as the rocket nears Didymos as well as Dimorphos.

What is the reason behind NASA colliding with an object in space?

NASA doesn't have to burn through $324 million to destroy the rocket without any particular reason. The agency is simply doing its work. It was in 2005 that Congress instructed NASA to discover 90% of the near Earth space rocks large enough to destroy a cityones that measure at least 460 feet in the width. This goal was set to be completed by 2020, however Congress did not provide NASA the funds to carry out the mission, so it remained the larger portion unfinished, and there are around 15,000 additional space rocks the same size still to be found.

While the office searches the sky for potentially deadly space rocks, they are also preparing strategies for what to do in the event that it discovers one at a crossroads with Earth.

This is the reason why the DART mission comes into play The mission is to demonstrate how hitting a surface rock by shooting it may cause it to move into an alternate circle. If you're aiming at a potentially dangerous space rock, this bump might be enough to shift the direction of the rock from the immediate hit to a miss.

1200 lbs DART rocket was launched in November last year. Imagine it as a fridge-sized shot with an objective that is 500 feet in width. The backlash effect can give power to the objectivethat is, the space rock Dimorphos -- and this provides it with a force which will alter its circular shape around a larger area rock called Didymos.

DART is experimenting with both the technology and innovation that is expected to hit a tiny target at high speeds -- a significant portion of the calculations rely on the calculations made in the direction of rockets, which also attempt to hit small targets at high speeds -and the ability to comprehend what happens to the space rock after it's hit. The release of garbage from the outer layer

Didymos will increase the force of that the impact.

Chapter 4: Colliding Dart Space Rock

NASA's DART rocket successfully collided with its space rock around 00.14 BST on 27 September. Here's the beginning and ending. It is important to be aware of this first-of-its-kind mission.

What's the point to this DART mission?
DART, also known as it's the Double Asteroid Redirection Test, is the first mission that aims to change the direction that a rock in space would travel smashing a shuttle into it.

This mission is vital for NASA's planet guard system and is a way to increase our ability to detect how we can foresee, predict and prepare for any space rock that could be an imminent threat to Earth. Is it a good idea to see one located?

It was launched via the SpaceX Hawk 9 rocket from Vandenberg Space Power Base in California on November 23, 2021. The rocket was damaged on the 27th September 2022.

"It's an important science experiment to determine if smashing a rocket into a space rock can be an effective way to switch its orbit around towards the Sun and perhaps diverting the Earth-crossing space rock in the near future should this occur, or better, should it happen," said cosmochemist and creator of the Shooting star Tim Gregory. Tim Gregory.

"It seems impossible to imagine for something that light as a rocket, or even the DART spacecraft that weighs the largest portion of a tonne, can hit something similar to the space rock, which measures a large amount of tonnes. But, you don't need to force the space rock through the largest amount to be completely off the Earth. It's just a matter of fractions of a degree or less, and you'll miss Earth by quite a few miles."

What rock in space did Dash meet with?
DART's goal was to construct to create a double space rock framework made up of a larger space rock dubbed Didymos which can be described as Greek meaning 'twin' and a smaller space rock called Dimorphos and is Greek for "two structures," which revolves

around it like clockwork. Didymos measures around 780m in size and Dimorphos is approximately 160m across.

"It's essential to ensure that this particular space rock structure doesn't pose any danger to the Earth. It was merely chosen as a possible target for this trial of science, and it was also viewed by a few participants due to its orbit around the Sun," said Gregory.

The rocket crashed into Dimorphos as it was about 11 million kilometers away from Earth. At the moment, it is believed to be travelling at 6.6km/s.

What's ready for DART?
It is powered by NEXT the an initiative of NASA's Developmental The Xenon Engine Business, a sun-based particle drive framework that creates push using the use of xenon as a fuel.

The camera on board is the high-end camera DRACO that is Didymos observation and Space rock Camera used for optical Route. In addition to being used to aid in the route the spacecraft, the camera was utilized to determine the size and condition of the focus on space rocks to determine the location of the site of the impact. The images captured by DRACO prior to the motor effects were sent back to Earth in a continuous manner.

In the final four hours prior to the influence, Savvy Nav which is also known as Little body Moving Independent Ongoing Route was close to DRACO to move the rocket to its position for influence.

The rocket also carried the same CubeSat known as LICIACube (Light Italian CubeSat for Imaging of Space rocks) which was designed in collaboration with ASSI. Agenzia Spaziale Italiana (ASI). LICIACube was launched on the 11th of September, during the rocket's journey to fight Dimorphos and captured images of the effects.

What happens following the crash?
After the event of the motor, the DART examiners will examine the effects of the crash of the rocket using Dimorphos by observing the ground with telescopes that have modern virtual experiences they've previously conducted. In this way, they'll

examine the efficacy of the motor's effect and determine the most effective method to use it in the event that future planetary security issues arise.

"The idea behind any type of research is often that you do not know of what's going on," said Gregory.

"Also given that this mission is as the very first one of this kind I believe that the potential for advancement is quite large. In addition, I believe that the mission will reward the Apollo space pioneers, hopefully it'll be a huge success, however, it could also be a disappointing one."

Chapter 5: The Effects At Chelyabinsk

The Sun was rising on Chelyabinsk. When the inhabitants in this Russian city headed off to work in the early morning of Friday, February 15th 2013 it was to be like every other day. As they looked at the dashboard camera, a practice that has grown popular across Russia Many of them took pictures of their journeys. But the cameras, typically employed to record minor traffic incidents could soon capture a once-in-a lifetime event...

In the beginning, it was unnoticeable that an unnoticeable speck of light struck the shroud of morning. In the same instant it appeared, the light speck was enveloped by a huge flameball that drowned out the Sun before exploding in the form of 25 Hiroshima Atom bombs. Just a few minutes later the shockwave destroyed the windows and doors for miles; the shards of glass harmed 1500 people. The town of Chelyabinsk was just hit by the strongest meteorite strike in more than 100 years.

While it is hardly an issue to many people, the incident was a stark reminder the extent to which our planet is. Although space is vast while the Earth is tiny There are numerous risks that are out there. It just takes one to alter the direction of life on this planet for ever...

In these photos we can observe that the object sank at a relatively small degree (top right) before its brightness rapidly rose (bottom left) and indicates the time the object started to break apart and let go of its energy. There was then an immense growth in luminosity (bottom right) after the object had completely disappeared. After the disintegration, tiny lights were visible as the individual pieces were sucked into the air.

While human lives are small to keep however, our Earth has a lengthy and varied history of such incidents. Relevant questions we may ask include:

• Where does it originate from?

* Why didn't you see it at the time?

What time will it be before the next hit?

How can we help protect ourselves?

In recent years, scientists from all over the globe have studied the immediate impact on the environment, their causes and the aftermath of these cosmic blasts. For the residents of Chelyabinsk the impact has given the birth to an exciting new winter activity looking for any fragments that survived.

fragments of alien rock particularly fresh ones are highly valuable. Naturally people saw an opportunity to make money by selling them on the internet to make a profit.

It's amazing that nobody in the world of science saw it coming. The largest object to strike the Earth for over a century came totally unannounced. Fortunately, thanks to the videos that people immediately uploaded to Youtube the entire duration of its journey through the atmosphere, from the moment that it landed, until its final fiery explosion, was captured on film.

While Chelyabinsk and the region around it was the most affected by the shockwave, due to the strength of the impact that it was detected across the globe. Monitoring stations across Alaska to Antarctica observed some of the biggest infra-sonic waves to ever be detected. The stations were not built for impacts with asteroid bodies or fireballs such as this, but rather for nuclear explosions, but they received the pressure waves that resulted from the impact, which as its power.

With the gathered data it has been possible to learn a lot about this meteorite. Including the force of the wave that caused pressure it released energy is believed to be approximately 500 kilotons which is around 25 times the force for that of the Hiroshima bomb. Examining this video, which shows the object was recorded from a variety of angles, suggests that the velocity of its descent was around 17.5 to 18 km/sec.

The energy a object absorbs is the result of its speed and mass and, since we know the first two above, you can calculate the third. In this instance it was about 7000 tons. From this mass, we estimate the size of around 20m in diameter. Large objects strike the Earth perhaps once or twice every century. The explosion occurred about 23 km above the earth, a lot of fragments made it all up to the top. fragments fell across a vast area of that was more than 100 kilometers long.

2. They were sent via the air...
Meteors as large as this aren't often seen so it's remarkable that scientists have complete information about the nature of meteorites, as well as from where they came. While the meteorite exploded in a massive bright, fiery explosion in the full image from Chelyabinsk the dashboard camera, a lot of its existence was in a completely different location and was given an entirely different name.

The life of a meteorite starts in deep space, being a tiny portion of a much larger structure known as an Asteroid. They can vary in size from a few meters across to over 1000 km. They are remnants of the nebula, from which the Solar System formed some 4.6 billion years ago. A large number of them are surrounded by in orbit around Sun in a circle known as the Asteroid Belt. When asteroids meet, tiny pieces split off. When they fall to Earth they can form two different shapes: the smaller fragments that explode into the atmospheric air are known as meteors. Pieces of debris that are large enough to travel across Earth's surface are referred to as meteorites.

Small particles of carbon dioxide are burned into the air.

While it's far from empty space Rocks, pebbles and dust are scattered across all over the Solar System. The Earth expands by 40 to 60 million tonnes every year, purely from the materials it accumulates. Before they reach space, the particles are called meteoroids. They are mostly tiny bits of space dust, and are without being noticed, however larger pieces, which can be many metres in diameter, do come frequently. The material isn't evenly distributed across space. Earth is swept up in different amounts throughout an entire year.

As it travels through miles of air every minute, frictional heat melts and melts the dust, resulting in brilliant streaks of light across the sky at night. Dust particles aren't large enough to be seen, however, larger pebble-sized pieces remain in the atmosphere, causing bright streaks of light in the sky at night. The streaks are known as shooting stars because of their appearance, but of obviously, they have nothing to do with have anything to do with stars. Meteors appear during the daylight hours too, but are snuffed out from the Sun. The larger chunks of meteors that measure centimetres (or even meters) across can generate stunning light displays as they drop. They can be amazing fireballs that shine in the sky. Sometimes, they are so powerful that they can be heard.

Thunderstones and fireballs The study of meteorites

Meteorites are long-standingly recognized as celestial objects. Records of meteorites, meteorites and fireballs that appear in the sky date way back to the beginning of time. According to the New Testament of the

Christian Bible, Acts 19:35, is a reference to a temple of Artemis inside which is the "sacred rock that fell out of the heavens". Chinese astronomers have recorded meteor showers as far in the year 700 BC as meteoric iron was found in jewelry in several Egyptian pyramids, including one of Tutankhamen.

The oldest meteorite ever believed to be falling and that is still present, can be found in the Nogata meteorite which struck the Suga Jinja Shinto shrine in Japan on the 19th of May 861 AD. The priests kept the meteorite in a container on where the day of its landing was recorded and, as an important precious object, it is kept in the shrine until today.

The first meteorite for that is documented can be identified as the Ensisheim meteorite that was spotted on the village in France (though being part of Austria in the year) in 1492. The sight of a huge fireball which was then followed by a massive explosion. After the dust dissipated and the smoke was gone, a massive 127kg rock that was about the size of a football was discovered in the middle of the village. Although it was not fully understood the significance of it, its value was in no doubt. The the Emperor Maximillian demanded that the stone to be protected in the town hall in the village, it is still visible in the present.

Over the next three centuries, several hundred of meteorites were observed to fall, and later recovered. In the early days they were not thought to be of any scientific significance. The first real scientific research into meteorites, also known as meteoritics, started in 1768, when a rock was discovered in the French town of Luce. Abbot Charles Bachelay, when presented with the rock, gathered several eyewitness accounts about the fall. He also produced a thorough report that was then gave in front of the French Academy of Sciences. A study of the stone and report was arranged by the famous French scientist Antoine Lavoisier.

The committee conducted the first ever chemical analysis of meteorite. The iron sulphide was identified within the stone it, they concluded that this stone was a normal piece of Pyrite (known through its formula chemically FeS_2 and also by the title "fool's gold"). It was left to the witnesses' accounts of its falling and it's black crust (created by melting during the process of falling) to clarify. It was believed that the blackish crust as they determined was created by the stone being struck by lightning. Thus, these stones became popularly known as "thunderstones".

The explanation of eyewitness accounts of the falling was straightforward the witnesses were peasants, so obviously they couldn't be believed. Without eyewitnesses of high-ranking to support the evidence the accounts appeared to be untrue and were therefore dismissed. Because no other theory of science could be developed and the thunderstones hypothesis was popular among the elite of Europe. It wasn't until a few years that scientists started to recognize the rocks as unusual.

But in the last 10 months of 18th Century and in the early years of the nextone, meteorite fall after meteorite fall was frequently discovered; it was impossible to let this phenomenon be neglected. In 1794 the German scientist Ernst Chladni, published a brief study of meteorites. He wrote on the characteristics of various iron meteorites as well on fireballs and the falling of many of these objects. He concluded that such objects originated from the Earth and were initially ridiculed however, it was gradually accepted over time and following further studies.

Large meteorites that were observed to fall in Siena, Italy in 1794 and in Wold Cottage, Yorkshire, England, a year later, sparked a lot of curiosity. The meteorites that fell in Italy as well as the discovery of a rock weighing 25kg in England that was later on displayed in London and resulted in the president of the Royal Society, Sir Joseph Banks providing samples to chemical scientist Edward Howard, who discovered the presence of iron and nickel that was not found on terrestrial rock. He also concluded that they could be extraterrestrial.

Many European scientists reanalyzed their samples and a meteor shower that was spectacular that was greater than 3000 objects in L'Aigle, France in 1803 eliminated any remaining doubts as to their origin and nature. Where they came from was not clear and some believed their source could be in the top regions of the Earth's atmosphere. Others claimed they were expelled by volcanic eruptions that hit the Moon. Chladni's assertion that meteorites came from far beyond the Earth-Moon axis and came from space was not supported by any evidence.

Additional analyses conducted during the 19th Century found significant differences between meteorites. The chemistries and textures varied like the minerals contained within. It was evident that meteorites might not have all come from the same source as the Moon. Recent advances in astronomy, specifically in the development of telescopes, proved it was highly unlikely that the Moon was home to active volcanoes on its surface. Additionally, it was found that meteorites travelled at extremely high speeds (many kilometers per second) which were far enough to be believed to have a lunar origin. In addition finding an asteroid around the year 1801 and subsequent discoveries resulted in the hypothesis that these fragments could be fragments of the planet. It seemed plausible that some of the pieces might have come into our path and hit on the Earth with meteorites.

In the early twenty-first century it became known that meteorites originated from space, however the exact location of these bodies was undetermined until around 50 years later. Some believed that they could originate from outside of from the Solar System entirely, but this idea did not match with the precise records of meteorite parent bodies that showed that they originated from within. Today, we can recognize meteorites for what they really are very rare glimpses into the beginning of the Solar System. Through them, we are able to examine the formation and birth of our Solar System, and the planets within it. The impact's role in shaping and shaping moons and planets are stored within these tiny bodies. by removing the outer layers, we are able to access an account of the processes that took place before even our Solar System was even born.

3. Where do meteorites come from?
Every year, millions of meteorites fall on Earth. They are recovered from every continent so it's not surprising that certain continents have yielded more than others. Environments that are favorable might have a large number of meteorites, when they are protected by the forces of nature. A number of thousands of meteorites have been discovered over 1000 were discovered to be falling. Sometimes, they are named in honor of the location that they fell, groups that are found within the same region are identified by numbers as well as an inscription, possibly

the most famous can be described as Allan Hills 84001, an object to be discussed later.

Meteorite Falls

Meteorites that are seen to fall and later recovered are referred to as "falls" while meteorites that are discovered and not observed as falling, can be classified in the category of "discovered". In 2013, around 40 000 meteorites were discovered and 1,100 were seen to fall. They fall roughly evenly across the globe but with the exception of the poles. Since comets, planets and asteroids within the Solar System are laid out in the form of a disc and more of the material is likely to be spotted from the equatorial direction.

Meteorites are seen to fall more frequently in the morning hours than in the evening and afternoon. When the sun is at its highest when the Earth is in the same direction the sun is traveling (see

The morning is when the Earth is the same direction in which it is moving. However, in the evening, it's traveling in an opposite direction. Because the Earth is able to sweep up objects while it travels across space, it's normal for meteorites to are observed in the early morning hours than other times of the day. A similarity to this is that more flies hit the windshield in front of cars when it is moving, compared to the rear.

In addition, at midnight at midnight, the Earth is moving in a completely different direction to the direction it's moving. The only meteorites visible will be those taking over the Earth as they have low apparent speeds. At night the Earth's orbit and movement is all in one direction, and thus, the meteors appear particularly rapid in the night sky.

There are numerous variables that determine how, when and what kinds of meteorites can be found. Naturally it is the case that there's a tendency based on the density of population, which means that meteorites tend to be found in areas with high population as opposed to areas where people are scattered. Furthermore, the kind of meteorite found is biased by the weight of the metal. Heavy objects are more likely to be discovered and picked up by amateurs as opposed to ordinary rocks that can be easily overlooked or ignored completely. Of course, the meteorite's kind doesn't affect the likelihood that it is observed to fall from the sky. Rock meteorites rule those in the "fall category" approximately ninety percent of all meteorites that are observed to fall are rock, however metal meteorites are much more frequent when they are in the "discovered" category.

Fireballs are produced by massive objects that have a lot of energy, like the one seen over Ensisheim which is more visible in the night. However it is much easier to pinpoint where a meteorite has fallen in the daytime. Also, the seasonal variation in meteorite falls can be due to their ease or maybe difficultyof recapturing the debris. In the summer and spring, there are more people out in the open and this increases the chance of seeing a meteorite falling, and the amount of people who could assist in the hunt. In addition it is easier to find a meteorite in the barren plains or frozen lakes is more straightforward than finding one in a forest.

To increase the understanding and knowledge of meteorites dedicated cameras that were able to observe the night sky were developed during the 50s. The cameras recorded thousands of meteors as well as fireballs, sometimes they also gave the exact location the sites where meteorites have hit. For example, in 2007 an enormous fireball was spotted on cameras across South Western Australia. While calculations pointed towards the correct directionbut also an investigation team that was later found two fragments, which had a total weight of 324 grams. This was less than one stone's distance away from the location of the impact.

By analyzing the frequency and distribution of the falls and findings by analyzing the distribution and regularity of falls and finds, we can estimate the amount of meteorites that are discovered every year. The search for meteorites is obviously very selective. 70% of surface of Earth is covered by water and there is a clear tendency to locate in areas that are well-populated. Meteorites that land in areas without people and the majority of terrain on Earth is similar to that - will not be observed by anyone. However, it is believed that between thousands and possibly the size of a golf ball hit the Earth every year.

Fireballs that are big enough and land at the correct location, can be seen by large amounts of people. The availability of video cameras growing in number and CCTV becoming more and more commonplace this footage, when combined with eyewitness accounts, could be used to determine the velocity and the trajectory of the object as well as to determine a likely landing point.

The Chelyabinsk meteorite is a prime instance of this, and it is a great illustration of the benefits of video footage of this occasion.

What do you do if you discover meteorites

While thousands of meteorites hit the earth every year, very few see fireballs or falls. If you happen to witness an event like this there are specific observations you can take which will help scientific research. The sightings of a fireball could be reported to national and local society for astronomy. If you have a nearby university with an Astronomy department, they could be able to help and give you the necessary contact information. The areas of scientific interest are:

1. What time of the day did the object to the ground?

2. Which direction did it go?

3. The object could have been a ball of fire, to what length did it burn?

4. What was the brightness of the fireball (as when compared to star, Moon and Sun) Did it appear colored or form a shape?

5. Did the object make a trail at the top of the mountains? What was it likeand how long did it take?

6. Are there some sound effects? If yes, what were they like? (Major fireballs can create an

sonic boom, however because sound travels extremely slowly in comparison with light it could take a few seconds or minutes, for the sound to be heard. This will aid in determining the distance that the ball is.)

In the unlikely event that you might actually witness the remains of a meteorite There are a few things you can take care of:

1. How long did it take after the fall before was the object found?

2. Where did the object go? (coordinates on a map , if feasible)

3. What was the weight of the object?

4. What impact did the meteorite have in the soil? For instance, has the meteorite harmed or penetrated the soil? If so, how large was the gap?

Photos or drawings that are relevant can be extremely helpful. The ideal is that meteorites should not be handled. While they aren't dangerous or hot they can be easily coated by terrestrial matter, that can be harmful to research. Ideally, they should be handled in gloves and placed in an envelope like an old-fashioned sandwich bag.

Meteorite Finds

In the 1960s, not even 2,000 meteorites were ever heard of, and more than 40% of them were believed to be falling. Today, around 1,000 or so meteorites are discovered every year, and usually just 1% are fall. This is due to the fact that some deserted areas of the globe are now believed to be a good source of food for meteorite hunter. Every year, many scientific societies organize trips to both cold and hot locations around the globe to search for meteorites. The ideal setting is one that is unchanging for long time periods and in which there is very little or no weathering , which could degrade the meteorites that are found there. There are two typical situations that can meet this criteria:

The Antarctic
The first meteorite to be discovered on the continent was discovered in 1912; in the following 50 years, several more were discovered in various places that were not unusual, considering that meteorites are found all over the globe.
In the year 1969 Japanese researchers discovered a vast meteorites-rich area in the blue ice fields on the south-western tip of the earth. In the past, only around a dozen or so

meteorites were discovered on every year across the globe. It's not surprising that the majority of them fell in areas that were populated, but numerous others were, and possibly were, and probably are, considered regular rocks that were never given the chance to be looked at. What is the number of meteorites that have disappeared into the forest, dumped in sand, or dumped down to sea level? In the forty years following the Japanese expedition there were tens of thousands of more that were taken from the ice. more than two-thirds of the meteorites in the world have been discovered in Antarctica. The most successful harvests have been found at the Allan Hills and the Yamato mountains. A lot of the rocks come from just a few meteorites. However, there are fragments from thousands of other meteorites.

The most obvious question is: why? What is it that makes Antarctic the best place to look for meteorites? There are many reasons. The Antarctic is not only the coldest region on Earth but it's also the most dry. This is a perfect climate for preserving and encapsulating meteorites over thousands, or even millions of years. The majority of meteorites found elsewhere on Earth were discovered in the last 200 years. Furthermore, because the Antarctic is mostly white in the color, finding dark rocks isn't too difficult. In areas where there are any other rocks, every similar object that is found in the ground is likely to be meteorites.

The glaciers in the Antarctic is not stationary. Ice slowly flows off the pole towards the ocean, bringing all the rocks along with it. But, if the path gets blocked by mountain ranges then all movement grinds to a stop. As the winds get stronger, they begin to peel away the top layer and reveal meteorites (as as well as other rocks) which were encased within. The regions that experience this are called "blue ice" because of the pale color they show. Satellites are extremely useful in discovering such regions, and identify areas that are suitable that scientists can explore for additional meteorites.

In accordance with the 1961 Antarctic Treaty signed by many large powers, such as those of the USA as well as the USSR The exploration of Antarctica is not permitted to conduct military tests mining, financial gain or other purposes. It is forbidden to make use of samples of geology for purposes other than scientific. Thus, all meteorites retrieved from this area are kept in institutions of science and cannot be sold or bought.

Baked Desert

Similar to the reasons of the cold tundra hot deserts are excellent to preserve meteorites. While the rocks aren't scattered and consolidated like they are at South Pole, the very dry conditions of the desert can limit the rusting and weathering of meteorites, thereby increasing their chance of being saved. But, huge dunes of sand that are blown by the wind are able to cover meteorites, and the flat, barren areas that remain unchanged for thousands or hundreds of years are great areas to explore. These areas are where the number of meteorites increases as time passes. In the Nullabor Plain located in South Western Australia has been similar to this for over more than 30,000 years.

In addition, during their descent through the atmospheric layer, meteorites are typically darker when their crusts melt due to the heat. They differ from earth-based, light-coloured rocks that belong in the desert.

There are areas that can contain hundreds of meteorites in just the space of a few square kilometers. They are usually result of a single meteorite breaking down into smaller pieces, resulting in an avalanche of stones across the area of. In contrast to Antarctica meteorites in deserts are able to be purchased or sold in a legal manner; however, certain countries, like Australia or Oman have laws that prohibit the sale and commercial use of meteorites.

Although they can be found all over Earth meteorites are difficult to identify after processing that makes them look like every other type of rock. From small pebbles to large chunks of boulder size, the biggest recognized meteorite has been the Hoba meteorite found on the farm of Namibia, South-West Africa in 1920. Around 60 tons of mass and 84% of it is iron, it has not been removed from the spot the spot where it was found.

It has since been made an National Monument with many visitors every year.

When you consider how much material is thrown at the Earth It is quite odd that there aren't many reports of meteorites striking people or buildings. There are a few instances of things or people who were in the wrong spot at the wrong time. in 1911, according to legend the most tragically Egyptian pet was killed. The first incident that has been reported to an individual occurred in 1954, when a 4 kilogram lump smashed through the ceiling of a house owned by a woman located in Alabama, USA; in 1992, an Ugandan boy was hit in the head. Amazingly, no one is believed to have suffered serious injury although there were some close encounters. Also, in the year 1992 another meteorite created an enormous fireball across the eastern part of the USA before colliding with the back of a Chevrolet Malibu in Peekskill, New York. While it could have been an unfortunate experience for the car's owner, both the meteorite and the car were later placed in museums due to their worth!

The Peekskill meteorite, captured on film as it slid above New York.

Illustration 3.4: The approaching of Peekskill meteorite. Peekskill meteorite in the year 1992.

A lot more have been found and put into collections or sold in pieces. From a scientific point of view the true value of a meteorite is that it functions as an "poor mankind's probe into space" as they provide information something about our Solar System without us having to travel to it.

Figure 3.5: A piece from the Barwell meteorite, which landed in Leicestershire on the 24th of December 1965. It was observed along the length of England the meteorite broke up as it fell; it might have been weighing as early as 46 kilograms.

Meteorites are shaped by the way they passed through the air. Sometimes the surface of the is curved or conical, signalling that the side in front of the drop and was the one to be heated the most. If the meteorite rolled over when falling, the meteorite could have a different form.

Chapter 6: Meteorite Composition And Their Origins

Meteorites are natural structures made of either metal, stone or a combination of both that have survived a space fall to the Earth. The majority of them are asteroids but a few originate directly from Moon or Mars and some even be comets. Meteorites are the earliest objects we can look at, and date to the creation in our Solar System, about 4.57 billion years ago. In this way, they offer us the opportunity to look at the past, and understanding the processes that formed this Solar System, and the planet that surrounds us.

Meteorites are classified into three fundamental categories based on the elements they're made up of. The majority (94 percent) comprise stony meteorites made up of quartz-based silicates (silicates) exactly like the ones discovered on Earth. Metal meteorites (5 5 percent) consist of the iron-nickel amalgam. Stony-metals (1 percent) as the name implies are composed of both.

Due to their metal composition, meteorites are heavier than terrestrial rocks. Here are a few examples of typical rocks found that are found on Earth:

Granite: 2650 - 2750 kg/m3

Limestone: 1550 - 2750 kg/m3

Chalk: 2230 kg/m3

Sandstone: 2220 kg/m3

Quartz: 2620 kg/m3

Ordinary Chondrites

Of the rocks, around 95% are chondrites. that are derived from Greek words "chondros" meaning "grain" also known as "seed". With the various pieces we have found, we've determined it is likely that Chelyabinsk meteorite was an chondrite.

Chondrites date back as far in age as Solar System. They were formed at the exact same time and from the same materials as Earth as well as the other inner planets. The study of them can reveal something about the early times of our Solar System, and how the bodies inside the system may be forming. They are a diverse mix of materials that include metal, but the most important constituents are tiny millimetre-sized chondrules.

Figure 4.1 The figure is a typical chondrite that can be found in Antarctica. The

chondrules are visible small circles, which are about a millimetre or centimetre in diameter.
Credit: NASA

Chondrules

Chondrules themselves are near-spherical millimetre-sized objects that once molten were formed at the beginning in the Solar System, and were gradually integrated into larger bodies such as the asteroids. They formed during the course of a rapid heating event, reaching temperatures that were higher than 1,400°C. The droplet of molten liquid then quickly was cooled, and then solidified. It remains the same way to this day. The precise mechanism by which this sudden heat erupted from is a long-standing issue within meteoritics (the study of meteorites). The answers to these questions will aid in understanding the initial process of planet formation and the accretion process of dust into larger smaller, centimetre-sized ball.

Metals

One thing that differentiates chondrites from terrestrial rocks can be seen in the tiny pieces of metal, typically nickel and iron. These make chondrites magnets which makes them a great method of determining an extra-terrestrial ancestry. The fact that metal is present in meteorites is extremely important and implies that the rock hasn't been destroyed since the early moments of our Solar System and, if it was not so the metal, which is heavier than rock, could be sunk into the center of the parent planet like it has occurred on Earth. In this scenario the pieces that fell off from the asteroid wouldn't contain any metal in any way. The amount and the exact kind of metal are a crucial way to categorize different kinds of chondrites.

Organics

Certain meteorites contain organic compounds, i.e. composed of carbon as well as oxygen, hydrogen and nitrogen. These are the essential elements that are required by living. Other meteorites may have come to Earth before life thrived here, and could have sparked life on Earth through the delivery of certain essential ingredients necessary for living.

For more than a century, it's been believed that meteorites could have transported organic molecules to Earth. Their discovery of organic compounds in newly discovered meteorites has led some to believe that they could be the result of contamination of earth's ecosystem. But, these suspicions were dispelled when, in the late 1970s an chondrite from Antarctica was discovered to contain various amino acids (important biological blocks composed of organic matter). After having sat in the Antarctic and in a clean environment and being collected by the use of sterile methods, no one could doubt the movement of organic matter through spaces.

Organic compounds, like amino acids, are available in two different forms, which are mirror images of one another. They are classified as either left-handed and right-handed (because they are able to rotate in one direction while in another) biological organisms that live on Earth exclusively use left-handed acids, for reasons that remain unclear. Chondrites however (no joke intended) contain identical amounts of both kinds which proves conclusively that the organic matter contained within them is not of terrestrial its origin.

This suggests it is likely that amino acids, as well as other organic compounds likely originated within the Solar System before living creatures were discovered on Earth. But, they aren't believed to be remnants of extraterrestrial life, although similar organic compounds found from meteorites in billions of years ago could be the precursors of existence on Earth.

Melted Meteorites

The normal chondrites we have discussed have seen only minor modifications since their formation; they have not experienced any significant heat or aqueous alteration. Some meteorites originate from bodies that have changed significantly throughout their lives, as a result of asteroids that were extremely heated (typically more than 1,000 to oC) and completely or partially melting. If a body melts it, the material that is denser sinks. For asteroids, the material is able to sink into the core, along with the rocky material around it. When the material breaks apart like this we can say that the body has been separated.

Figure 4.2 The process of differentiation. The process by which an object such as a planet an asteroid is transformed from a body that has material evenly distributed and an identifiable core, mantle and crust, based on the density differences of the material of the body.

1. A primitive body that has an overall spread of material before the heating process begins. (top left)

2. If the body melts, the heavier substance sinks into the middle, while the lighter materials floats on the surface. (top right)

3. The body has been completely split into mantle, core and crust. This has happened across all planets of the outer Solar System, Earth's Moon and on the biggest asteroids. (bottom)

There are three kinds of melt meteorites: metals stony-metals , and achondrites. Each is derived from different segments of a distinct asteroid.

* Metals

Metal meteorites are usually composed of iron and varying quantities of nickel. They are formed from the core of different asteroids.

* Stony-metals

Stony-metals as the name suggests are a mix of two different materials. They are formed from the border between the core of the metal and the outer regions of rocky rocks. Some are believed to have formed after two asteroids of different sizes met.

* Achondrites

We've previously discussed the chondrites of rocky meteorites, which contain the chondrules. The majority of the rocky meteorites contain chondrites, while the rest are called achondrites and they don't have chondrules. The textures of meteorites are similar to the igneous rocks. This means that they were heated up and melted upon formation - thus eliminating Chondrils - then they cooled and crystallized once more, indicating that their mother body was volcanic in the course of the course of its life.

Lunar Meteorites

The best method to research an object is by collecting specimens of the object. If we want to study the planets and asteroids, we could visit them, take photos with spacecraft or even crash on them and take samples. Space missions, however, are very expensive (bidding begins at around $200 million and can go up to) and are also extremely very rare. On Earth however, meteorites from these bodies can be found all over the globe. All we need to do is search for them.

While the majority of meteorites are derived from asteroids. A few could be derived from other bodies within the Solar System. The the only other sources we have knowledge about are those of the Moon and Mars however there are some that are thought to come from comets.

Our closest celestial neighbor is the Moon is about quarter million miles (400 000 kilometers). This is the one other body that the Earth that humans have walked. This is why we have more knowledge concerning our understanding of the Moon as compared to any other celestial body on the planet. Astronauts have returned stones from space missions. We also have evidence of the Moon which have arrived on our planet on their own volition, and are known as meteorites.

We've known for a long time (or at the very least, believed) that meteorites originate directly from the Moon. In the past there was a belief that all meteorites originated directly from the Moon. Asteroids were discovered around the beginning in the 18th Century suggested a different source, a notion that was slowly accepted. Recently, cameras specifically focused on the sky have revealed that the majority of meteorites' orbits originate from asteroids, they are not derived from the Moon. A strange meteorite was discovered in The Allan Hills in Antarctica. The small piece of rock (ALH 81005) is an opaque dark grey clump with some white patches that are just several centimetres in size. If you examine it, it becomes evident that it was like other rocks returned out of on the Moon from the Moon by Apollo astronauts. The majority of the lunar meteorites we have are found in Antarctica or in different African countries, particularly within the Sahara Desert. Some have been discovered in Australia.

Figure 4.3 Figure 4.3: The Moon is the source of more than 80 meteorites captured with Galileo satellite. Galileo satellite.

Credit: NASA/JPL

When we look up at the Moon We can see both light and dark shades. The regions with creamy colours are the highlands, while the dark regions are the lowlands. The two regions differ more than altitude, the highlands and lowlands are composed of various minerals, which create the shades of the meteorites. The meteorites which are blown off the lunar surface after they were struck can originate from all areas of the Moon and thus show different types of surfaces and compositions. A small portion of the moon's surface Moon has been studied by scientists, and meteorites may provide useful data about the regions that remain to be explored. Certain meteorites originate from the outermost reaches of the Moon which is a considerable distance from any previous landing locations. Spacecrafts' visits to certain areas are not always able to provide any clear view of the Moon in general and any information which can help in this regard is very welcome.

Martian Meteorites

There are currently over 100 Martian meteorites that we have in our possession. The majority of them are young in comparison less than one billion years old. If they were formed relatively recent, it is possible to conclude that they were formed from a body which was, up until a while ago geologically active. The asteroids all cooled around time ago, around 4 billion due to the fact that they were tiny and didn't have enough mass to sustain temperatures that were high inside and therefore, they could not be from any asteroids.

Because they are mostly an igneous (volcanic) rocks that means they have a planet-based source (or perhaps a moon that is large) because the temperature required to create an eruption can only be generated within the center of a planet, or another large body. The proportions of certain of the elements found within them, including oxygen, xenon, and argon, differed from Earth's, suggesting a world with an atmosphere that is thin. In addition, the ratios of these elements was similar to (or comparable to) the data from the Viking landers who landed on Mars.

Martian meteorites also referred to as SNCs They are also known as SNCs Shergotty and Nakhla -- Chassigny because of the places located in India, Egypt and France in which the three meteorites were first discovered (in 1865 1915, 1815 and 1865) although they weren't recognized as Martian at first.

Life on Mars

Meteorites that originate from Mars are particularly interesting due to the fact that their home was believed to be home to life, possibly even intelligent life, on the surface. Mars is the planet of the Solar System most like the Earth and is a terrestrial planet in its composition It is a mountain with mountains and valleys There is evidence that water flowed on its surface. There are close-up images of the surface as well as decades of space rover research has given us a good understanding of the geology, geography and the atmosphere that the planet has. The water is frozen in the poles (though astonishingly, even the fiercely scorching

Mercury has frozen the water near its poles) However, our fascination of the world is due to a issue that is still unanswered If Mars is like the Earth is there ever been any life on Mars also?

A tense and perplexing answer, a clear "yes" could prove this phenomenon isn't unique to Earth. The existence of life on Earth within the Solar System could just be an isolated, lucky roll of the dice. However, if life developed on two different planets that were not dependent upon one another, and especially inside the exact same constellation it is unlikely to be an unplanned event and the Galaxy and the Universe is surely bursting with life.

The idea of living at Mars has been around as long as telescopes which it first became visible in full detail. In the late 19th century, telescopes were sufficient to allow astronomers to draw basic maps of features on Mars. Light and dark patches could be observed and thought to be forest and seas. One astronomer Giovanni Schiaparelli created detailed maps that he used to draw strange, artificial lines were visible across the deserts that were light-colored.

The other, Percival Lowell, who built a huge observatory in Flagstaff in Arizona specifically to study Mars and the moon, believed that strange lines that ran across deserts of ochre were the result of intelligent living things.

"...That Mars has been filled with beings of some kind or another, we can think of as a certainty, but it is not clear what those beings might be. ..."

- Percival Lowell

Figure 4.4 Schiaparelli's chart of Mars with numerous canals, that have since been proven to be inexistent.

If existence did occur on Mars What would it appear like? A lot of sci-fi writers let their imaginations run wild with this issue. In the novel The War of the Worlds, HG Wells described a Martian invasion by massive strange creatures, in huge cylinders that were mounted on tripods that burn everything down by emitting heat radiations! In the days of television, those in the props department probably enjoyed creating anything they could come up with However, once science was applied to the concepts, they have turned out to be completely different.

Figure 4.5: The attack of an Martian tripod Original image taken from "The War of the

Worlds" by H. G. Wells. These ideas are incredibly thematic and, since in the early 19th Century (and most in the latter half of 20th Century) Mars was unknown and unexplored, these ideas resonated with a lot of people.

Some theories suggest that because harmful ultraviolet radiation is able to penetrate the surface of the planet, life forms are likely to have strong shells that keep them safe from this. Additionally, Mars is known to be extremely dry, so any living thing must concentrate on the protection of water. As an example, it may be plants with "fine roots-like probes seeking, not fluid water but for permafrost in the deepest parts of the earth to locate the frozen water. ..."

Biologists were less imaginative rather, they conducted studies which tested for microorganisms that might have grown on Mars. There was a belief that if they lay dormant, a mix of nutrients could re-energize them. But, of course, those theories as well as others were only guesses. The only way we can comprehend life would be to travel to the planet. But even the strongest telescopes were not enough for the planet. It is too tiny and distant. At the end of the 1950s, spacecraft that were not manned had made it to the Moon however Mars was over 1,000 times further away to reach; the process takes more than a year therefore even the smallest error could see the spacecraft stranded.

The first probe that was truly successful that was successful Mariner 4, which after traveling more than one million miles per day for eight months came to rest in July 1965 and sent back 22 images of its surface. A complete departure than what one would expect the surface was dotted with craters. Many of which had developed a long time in the past. This suggests that there was not much weathering on the surface or geological activity generally. The surface wasn't like the Earth in any way It was similar to the Moon which suggests that it had remained in a state of pristine condition for billions of centuries. Also, in other words: Mars was dead.

Two years later, another craft were launched with an even more powerful camera than the one they had before. Although Mars could appear like that of the Moon in a distant view the hope was that the more detailed images of the Moon, taken from a close distance, might provide a new and exciting view.

Patrick Moore, broadcasting live on the BBC was swept by the excitement

"...We've recently received some incredible photos returned from the American mission that was sent to Mars, Mariner 6. Look at that the craters that are on Mars like the Moon! The biggest crater in that image is around 160 miles in size. Remember that when Mariner captured that image it was just one-third of the way as the surface of Mars like we're today from Moscow. I'm curious about how the holes got there, and what they are? Could they be due to something striking Mars and if so, are they volcanic? I am convinced that the majority these are volcanic but I'm yet to be proven to be wrong. ..."

"...You can see some dark areas that may be plants and at the bottom, you will observe the clear polar cap, which has been believed to be a type of frozen or icy layer. ..."

Figure 4.6 The first close-up images of Mars were taken in Mariner 6.

Credit: NASA/JPL

However, examinations of the atmosphere showed that the atmosphere was extremely thin, mostly composed of carbon dioxide. Scientists were dismayed. In the meantime, the Apollo program for the Moon was enticing the general public. When Armstrong and Aldrin took a walk onto the Moon in 1969, it was at the time the most watched broadcast on television ever. In secret in it was revealed that the Soviet Union was preparing an even more ambitious plan to fly to the Moon. They succeeded. In 2001, Mars 3 landed safely and then, after 90 seconds, the spacecraft completely stopped transmitting!

NASA's Viking team had more success It was first launched in 1975, and was in the orbit of Mars without issues. The 10-month journey to Mars was not easy and space travel is rife with dangers including hard vacuum, lethal radiation and freezing cold, sweltering heat, and any solar flares, all of which could damage delicate electronics or even destroy the spacecraft completely. However, the last seven or so minutes were even more challenging: Viking had to survive a fiery dive through the atmosphere, slowing down between 21,000 kilometers per hour (5.8 kilometers/sec) to a halt when it reaches the surface.

"...And it's an initial piece of information that's coming into. It's a shame... You're supposed to be saying something in this moment, but I'm just not speaking! There are rocks. Yes, there are rocks. There are rocks, in the form of... The rocks are... Wow, it's amazing to realize that... Mars, it's really there. Then again... It's a quick interpretation can be a slightly risky, however many of these stones have a similar appearance to the ones we've observed previously in... the desert landscapes. ..."

JPL Caltech's Viking Centre upon the landing of their craft.

Figure 4.7 The rough surfaces of Mars.

Credit: The Viking Project, M. Dale-Bannister WU StL, NASA

The view was stunning, without doubt however the most important concern to be answered was whether there is existence on Mars. One week after landing, the robotic arm grasped and snatched the first piece of soil. The mini laboratories of the craft spun and whirred through the soil samples and sent the results back to many millions of miles. Scientists eagerly awaited the results in the hope that they would be just about to conclude more than three hundred years of speculation.

A machine that was on the lander, called a "gas chromatograph instrument" was able to analyze the gases released by the soil after it was heated by a tiny oven that was built onboard. After a couple of weeks of testing, the standard gases such as carbon dioxide, nitrogen and argon were discovered to be were trace amounts of water and oxygen. However, the molecules that indicate living organisms were not present. This was a complete surprise and some could not believe it and were skeptical about the sensitivities that the spectrometer had.

However, there was no explanation that could stand up to examination. While the elements to create life were in place however, life itself wasn't. However, these were exciting results and scientists were keen to go back to Mars to study further. However, the political system stepped in and stopped funding a second mission. It wasn't for another 17 years before another spacecraft made its way towards the planet of red. NASA launched many mission in the direction of Mars in the 90s, however, despite their best efforts the majority of them did not succeed. An especially embarrassing error afflicted that mission Mars Climate Orbiter, a mission that cost $325 million. Scientists couldn't understand why their spacecraft was lost on the horizon... up until the time they realized that everything was planned in inches and feet instead of centimetres or metres in the way that was intended!

In just two months, NASA had to admit another embarrassing error in which, for unknown reasons it was discovered that it was discovered that the Mars Polar Lander failed to reconnect with Earth following its landing. The challenges of spaceflight are evident that thousands of calculations and processes are carried out on a space mission and any one error could prove devastating. Discovering life on another planet was difficult at the most ideal of times however, the technical challenges in achieving the other planet made it appear unattainable. But, in the hunt for extraterrestrial life the breakthrough was made from a distance. Geologists have been looking for Martian meteorites with the hopes of learning more about the Earth and the benefits of meteorite exploration made the top of the news pages.

ALH 84001

Perhaps the most well-known of meteorites ALH 84001. Discovered in Allan Hills in Antarctica 1984 It was the first specimen from the expedition to be classified and processed (hence the number 001). It was greenish-colored and was believed to be through the Asteroid Belt however, the presence of carbonates, minerals and iron in it created it to be a rare meteorite in the world. The only way to find out was it later discovered to contain iron disulphide, and not mono-sulphide like one would expect and the latter was typical of SNCs and, as such, it was classified as Martian.

It is fascinating to note that at 4.5 billion years old, it's not only the most ancient Martian meteorite (more than double older than the following) and also dates back to an era in which Mars might have had large amounts of surface water. Carbonates found on Earth usually indicate the presence of life. Therefore, we were focusing on this possibility. Indeed under extreme magnification it appeared that fossilized bacteria were discovered!

First time in the history of mankind it was discovered evidence of life on another planet!

Figure 4.8 Image (about 1 mm to 10-6 meters across) from the scanning electron camera of the field of ALH 84001, which depicts an alleged microfossil (highlighted).

Credit: NASA

"...Today Rock 84001 has a message for us across trillions of years billions of kilometers... In the event that this finding is confirmed it will certainly be among the most astonishing discoveries of the Universe which science discovered. ..."

A speech excerpt delivered by the former US the president of that time, Bill Clinton.

Naturally, the news media turned into a flurry, however, the excitement didn't last long. It was unfortunate that they were so tiny - the "fossils" were only 20nm across and 50nm long and smaller than any other known living cell. Perhaps, the fragments could be a result of a mineral formation, or caused by contamination from terrestrial sources. While this and other rocks that have similar structures don't indicate that life was once present on Mars however, it does suggest the necessary conditions to sustain life were present on Mars some time ago.

The ingredients for the quality of your life

To allow life as we know it to be there have to be a number of prerequisites and ingredients to begin among them is water. What makes water such a useful ingredient is its capacity to contain and transport solid substances inside it. It is a basic compoundmade up of oxygen and hydrogen. It is usefuland retains a liquid form across a range of temperatures. Thus, the quest for life within this Solar System revolves around the searching for water. Those areas where liquid water is present or has at least existed, could be potential habitats for living things. It is widely believed that a large portion of the Earth's water came from comets that struck our planet. It can be found in a variety of Martian meteorites. However, a particular meteorite that was discovered two years ago has taken it to a whole new level.

Figure 4.9 Figure 4.9: The Martian meteorite Northwest Africa (NWA) 7034 is known as "Black Beauty" is a massive stone with a weight of around 11 pounds (320 grams). It was discovered within the Sahara Desert in 2011, this rock is a completely new type of meteorite. Analyzing it has revealed that it's more than 2.1 billion years old which makes it the second-oldest rock found on Mars and was formed at the initial phase of the current geologic phase on Mars called the Amazonian. It's a piece of a crucial period in the development of Mars.

What makes this rock intriguing is that it has 10 times the usual amount of water as compared the other Martian meteorites that have no known origins. The rock is composed of basalt, a mineral formed from rapidly cooling volcanic lava, which suggests earlier eruptions on Mars. This unique composition is in line with the Martian crust, as seen by NASA's surface-rovers and The Odyssey Orbiter.

Credit: NASA

Although Mars being comparable to Earth in terms of size as well as distance to the Sun The conditions and climate in the first are unsuitable for living as we know it. Mars is tiny, which means its gravity is not as high (38 percent of Earth) and a large portion of the atmosphere is escaping. With a lack of atmosphere as well as no layer of ozone there isn't much protection from ultraviolet radiation from the Sun. In addition temperatures are extremely cold, and the atmosphere pressure is very low (below one percent of Earth) that complex life forms (like humans) must keep from preventing your blood vessels from getting boiling.

In the past, Mars was quite different from what it is now. It was once a place where water flowed easily across the surface, suggesting that it was surrounded by a thick atmosphere that kept it liquid and prevent it from freezing. Atmospheric clouds are formed by volcanic eruptions. On Earth these eruptions continue till this day, but there was a time when on Mars there were fewer volcanoes and eruptions weren't as frequent.

There isn't any evidence that suggests life exists on Mars today or has previously existed. On Earth there are many kinds of life that thrive even in extreme environments. The creatures, collectively known as extremophiles, are found in the ocean floor living under extreme pressure (a couple of miles of water over you is quite a bit of weight) as well as in frigid temperatures, and with no sunlight. Based on tests that are suitable have shown that bacteria can live in space, regardless of the range of temperatures and radiation which suggests that life might exist on Mars or other moons and planets that have extreme conditions. If life ever did exist on Mars or is likely to ever be in the near future it is likely to look very different from what we see on Earth.

Chapter 7: The Asteroid Belt

Are asteroids a type of asteroid?

When when the Solar System formed, eight bodies that were in the disc around the Sun have grown large enough to control their respective regions. In addition to sweeping up material and gas from their path They cleared their roads of traffic. They've since been called the planets after the Greek term for "wandering Star".

But, in other parts of the Solar System there are vast reserves of not explored, cold rocks. Particularly between the orbits Mars and Jupiter and further than the orbit of Neptune and within an immense shell on the outermost portion of the Solar System, innumerable scraps from the early Solar System dwell.

The biggest part of the innermost group, Ceres was discovered accidentally at the beginning of the 19th century. The discoverer, Giuseppe Piazzi of Italy was working on the catalogue of new stars when he spotted a new star. He described it this way:

"...The the light seemed faint and in the shade of Jupiter however it was comparable to many other light sources that typically are regarded to be to be of 8th magnitude. Thus, I had no doubt of it being anything other than an unmoving star. At the end of the second, I revisited my observations and after observing that it was not in line with respect to time or distance to the zenith as the previous observation I began to think about doubts as to its precision...

...The night of the 3rd night, my suspicion was transformed into certainty after I was sure that it wasn't an unfixed star"

A couple of days after, he wrote to one of his friends telling him:

"I have made the announcement that this star is an asteroid, however, since it doesn't have any kind of nebulousness and, moreover because its motion is very slow and homogeneous, it has occurred to me numerous times that it could be more than being a comet. However, I've been cautious not to make this assertion to the general public."

Other astronomers were less skeptical about their theories and were convinced that this was a new planet. In the subsequent years three more "planets" came to light. Ceres wasn't the only one but was the biggest of the group. Astronomer William Herschel suggested they be known as asteroids, which comes taken from the Greek word for "star-like" which is why the name has remained until today. A fifth asteroids was discovered in 1945, 38 years later than the first however, in the years since, at the very least one asteroid has been discovered nearly every year.

Yet for an asteroid Ceres is a lot bigger than the rest and very like a planet. with a radius of less than 950 km, Ceres is just 0.01 percent of the mass of Earth but is over three times as large than the next one, Vesta, and the only one of them with enough mass to be able to collapse into a spherical-shaped shape.

There are currently hundreds of thousands of asteroids identified and formed an entire ring around the Sun that is known as the Asteroid Belt. Except for moons (which we'll discuss in a bit) the most tiny asteroids that are known are only hundreds of metres across There are millions of smaller pieces and an innumerable amount of pebbles. Yet, despite their sheer number however, the total mass of all asteroids is only 5% of the size of the Earth's Moon and they are greater in their spread than what is typically used in Hollywood films. The most well-known pieces are usually separated by several hundred thousand or millions of kilometers even though the surface indicates that they collide and break into pieces, we have not ever witnessed this happening.

In fact the asteroid belt is so thin that a number of satellites have passed over it undetected. Some of the most well-known spacecraft that have emerged from the vast size unharmed include:

* Pioneer(s) 10 and 11
* Voyager(s) 1 and 2
* Ulysses
* Galileo
* Cassini

Figure 5.1 Asteroids in the main belt. The distances are all shown in astronomical units. 1 AU represents an astronomical unit that measures the distance between The Earth to the Sun. Mars is an elliptical orbit. This means that it can be seen it is as close to 1.38 AU, and as far as 1.67 AU. Jupiter is a bit different in its orbit, but it doesn't go away from the 5 AU limit. Certain asteroids do not belong to the main belt, however, the majority of them are in the area. In terms of orbital durations: Earth obviously orbits every year, whereas Mars is nearly twice as time (1.88 an hour). Main belt asteroids can take anywhere from 3 and six years for orbits, whereas Jupiter completes its orbit about our Sun every 11.86 years.

Note An aside: The Density of belts has been increased to provide artistic effects.

The Origin of Asteroids

Two theories are for the origins of asteroids. According to the first theory, asteroids are remnants of a planet which was broken up. Another theory states that asteroids represent the fragments of the planet that failed to unite. The truth is believed to be a mixture of both theories.

What we know for certain is that the mass of all asteroids has less than of the Earth's Moon that is way insufficient for the size of a significant planet. This means that we can eliminate the origin theory that first came to mind. There is evidence to suggest that a variety of asteroids once gathered into several massive bodies that later split into pieces.

Why didn't the asteroids manage to unite into a single tiny planet?

As it was believed that the Solar System began to form around 4.6 billion years ago emergence of Jupiter and its massive gravity, drastically altered the local gravity forces. Its impact shook the motions of the smaller bodies that were nearby which caused them to be thrown into collision and then to fragment, ensuring that they didn't grow too massive. No planet could ever develop within itself or between it and Mars.

The planets that were in the inner solar system survived because the building blocks they used were far away than Jupiter. They continued to orbit in circular orbits, at an almost constant speed. They gradually merged and increased in size. In the end, four terrestrial planets merged - Mercury, Venus, Earth and Mars. If not for Jupiter and the outer SS might have had five members, but maybe a tiny one.

Asteroid Composition/ Structure

Very few asteroids have symmetry. Many have jagged or jarred edges where bits split off, and several craters that result from eons of collisions with other rocks. In the course of the Sun Most revolve around their axes at different speeds and in a manner that is quite irregular. There are upwards of 150 that have a smaller moon as a companion while some are greedy and have benefited themselves to two. There are also systems in which bodies of similar size orbit around a common center and there are other systems in which three asteroids have steady orbits, held to a certain degree by the two other.

Except for the rare collisions Asteroids have not changed much over the last billion years. Studying them can reveal many things about the beginning of the Solar System.

Asteroids can be classified into several kinds based on their composition. There are a variety of asteroids that are part of the central belt. The most common include:

* Carbonaceous (C-types),
* Silicates (S-types) and
* Metals (M-types)

The name implies that C-types consist of carbon, which makes them extremely black (like coal) and a little blue. C-types outnumber other asteroids, accounting 75 percent of the the materials that is found in the central belt. The largest asteroidis Ceres, Ceres is a great (and huge) instance of C type. The apparent spectrum of light appears flat and smooth and it has features of infrared absorption which are believed to be due to clay-like material that has water in.

Figure 5.2 1. Ceres is the biggest of the objects in the belt and was the first to be discovered (the number preceding with the names of asteroids indicates the date they were discovered and hence"1" for Ceres) "1" to denote Ceres). Ceres as seen by NASA's Hubble Space Telescope the sole one with enough mass to collapse gravitationally into the shape of a globe. This feature was the reason that led to its classification as dwarf planet. It indicates that the interior contains multiple layers, with the solid, rocky center beneath an icy mantle and an extremely dusty, thin crust. Around 950 km in length The variations in colour as well as brightness may be due to impact factors or caused by differences in surface material.

Credit: NASA/ ESA/ SWRI/ Cornell University/ University of Marylandand STSci

Around 15 percent of the asteroids in our universe are of the S-type. Some of them are reflective of sunlight in a manner which suggests a metallic-looking surface. Asteroids of different types tend to be found in specific distances from Sun. The outer asteroids comprise sand, silicaceous, and dry (S-types) carbonaceous (C-types) with clay-like and water-rich materials. They are found in the outer edges of the belt. M-types are the most prevalent within the middle.

It is possible they are the most bright asteroids because they lie closest to the Sun and the darkest are farther out and you're right. But, as they are composed of various materials their proximity to the Sun aren't the only thing that determines their luminosity. The asteroids located on the inner portion of the belt don't just get much more light than objects farther out, but they reflect more of it as well. The percentage in light that objects reflect its albedo. The range is between 0 and 1 an object that has zero albedo is unable to reflect any light at all; an albedo of one means that the object is reflective of all light that hits it. Different compositions and reflecting ability of asteroids may be caused by the conditions of the solar nebula, from which they originated.

Formation of Asteroids Formation of Asteroids

The asteroid belt itself serves as an excellent line that divides the inner and the outer Solar System. This isn't just because it divides the smaller terrestrial planets from the massive gaseous ones, but also because it defines the boundary between the condensing, cold regions and the hot, water-vapor part of the solar Nebula. The asteroids that are darker, and filled with water and carbon, can condense in cooler temperatures farther away from the Sun however not as close to the Sun in which water can be evaporated. Further away from the Sun the only objects that could be formed were solid bodies, which is why the dominant role played by these types (and types M) there.

Alongside the three kinds There are a variety of smaller sects with an array of shapes, colors and flavors (or at least , the three of them). There are many different types, but they only account for tiny fraction of the asteroids in between.

The measurement of the size of asteroids

Even with the finest telescopes on Earth asteroids are so tiny that they appear to be tiny specks of light, much like an unimportant star. While they might appear like stars but their physical properties is quite distinct. Utilizing instruments of science, we can find out that asteroids are smaller than the size of any star or the planets of major importance.

One method used to measure how big an asteroid's size is to observe the shadow it leaves as it passes before the star. When the Earth turns while the asteroids appear be moving across the sky, the observer on Earth can determine the size of the asteroid based on how long the star is unnoticed. This has been proven to be true on the biggest asteroids 1. Ceres as well as 2 Pallas.

Another method to estimate the size of an asteroid is measuring the apparent brightness. In general, the larger an asteroid, the greater amount of light it reflects which is why it appears. We've seen before that the composition of an asteroid directly affects the percentage of light it is able to reflect. For instance, 1 Ceres is by the most extent the largest asteroid within the belt. However, having an albedo of 0.09 It is actually 4 Vesta and the highest albedo, 0.42 which is the brightest (and also the one that is visible through an unaided eye).

Size and albedo can be determined by looking at infrared and visible wavelengths. When sunlight hits an asteroid, some could be reflected, but the majority will be absorbent. In the process of absorbing energy, objects begin to get hotter and release infrared light. When we measure the infrared light emanating from an asteroid, and comparison with the reflection of visible light we are able to estimate the albedo of an object. If we know the Asteroid's brightness and location from ours, it is possible to estimate the size of the object.

The smaller asteroids vastly surpass the larger ones. Around 1000 asteroids have 30km in size. Studies of asteroids with faint colors indicate that there are an additional half million that are more in diameter than 1.6km (1 mile) across.

Most asteroids revolve every 2.4 or 24 hours. As asteroids (with the exception of Ceres) aren't perfect spheres, when they rotate they produce different quantities of light, which could help in their identification. There are asteroids that require longer than others to turn, like 253 Mathilde that takes 17.4 days to turn. Some complete one revolution in a matter of hours or minutes. They are however very small. A large asteroids (upwards 100 metres across) is unable to spin at a high speed and remain in place all at once.

Although the majority of asteroids are located inside the central belt there are some that reside close to the Sun and are close to, and even traversing the orbit of the Earth. There are many variations between the sizes and status of Near-Earth Asteroids (NEAs) as in comparison to those inside the central belt (MBAs).

For example, NEAs have a much larger range of speeds for rotation than MBAs. Additionally, a higher percentage of NEAs are retrograde - they rotate in the opposite direction from the way they revolve around the Sun. Also, a greater percentage of NEAs are binary asteroids opposed those in the central belt. What is the reason for these strange behavior result?

It has been proven that the tidal (gravitational) interaction with bigger bodies could influence orbits and movement all over. As we'll discover, many Near Earth Objects were removed from the asteroid belt because of the impact of Jupiter. Many billions or millions of years of interaction with Earth and Venus can have significant impact on the spin and condition that the object. Certain objects will see their spin increase, whereas other slow down. When they are over their limits of rotation may be broken up, with pieces flying in different directions or, in certain instances, the pieces are able to come back together as a binary, not being a single object. This could explain the high proportion of binary asteroids found in Near-Earth orbits.

Figure 5.3 Formation of asteroids that are binary. If an asteroid spins too fast, may split into smaller pieces. The fragments could break away in different directions or they may remain connected by gravitational force and, in this case, they could reassemble to form a single, squishy asteroids. Or, they could make two (or more) asteroids that bind to one another and orbit the Sun in a binary.

The Movement of Asteroids and Jupiter's Command
Kirkwood Gaps - The dead zones
Nearly six decades after the discovery Ceres (by the time there were more than 50 asteroids known to exist) it was the American scientist Daniel Kirkwood noticed there were numerous empty orbital paths within the belt. They are now referred to as "Kirkwood Gaps" and represent evidence of local gravitational forces that govern this region of the Solar System.

At certain distances from Sun Asteroids have orbits which put them in a resonance with powerful Jupiter. That means that their orbits are exact fractions of Jupiter's orbital duration. For instance, a spacecraft which had an inverse 3:1 resonance with Jupiter would have completed three complete orbits for each single orbit of Jupiter. This kind of object would be around 2.5 AU away from the Sun and would take approximately 3.95 years to circle the Sun approximately 1/3 of Jupiter's duration which is 11.86 years. A asteroid that crossed this area would be in the exact places with respect to Jupiter every time it crossed. The constant tugs on an asteroid during these locations would eventually remove it from its orbit, and force it to a longer eccentric orbit, or even eject it from this region of the Solar System altogether.

Figure 5.4 A: When a tiny object, like an asteroid and a larger mass, similar to the planet, meet regularly the larger one repeatedly pulls on the smaller through its gravitational pull, ultimately taking the smaller body from its orbit completely.

Asteroids that don't occupy the clear resonances are still affected by Jupiter however, they'll approach in different areas of the orbit. A constant tug on the same spot on an orbit will have an impact on the whole orbit and nudges on a variety of places tend to cancel each other out and result in little or no impact.

Figure 5.5 Asteroids that are ripped from their circular orbits and into chaotic ones can travel anywhere. Certain orbits that are distorted bring these objects out of the belt, traversing the path, and even collisions with Mars, Earth, even Venus and Mercury Sometimes, they do not return back to the belt. Some will completely disappear completely from this part of the Solar System. It is important to note that these changes happen extremely slow, requiring thousands of orbits. The diagram is exaggerated for aesthetic reasons. Only a few bodies have been captured by the planets that orbit them and became moons. The planet Mars has captured two bodies: Phobos and Diemos, with a distance of just 27 and 15 km, respectively.

Different types of asteroids.

Over 90% of asteroids are in the main belt , which lies between Mars as well as Jupiter. They have been following their paths for millions perhaps billions, of years. In space they pose no threat to our planet, but occasionally an asteroid alters its direction. There are numerous orbital resonances that are associated with Jupiter. While objects might be able to enter them, they don't remain for long. The shifting of the orbits of asteroid bodies which caused them to collision with each other and prevent an additional planet from forming within the area between Mars with Jupiter.

Figure 5.6 The Distribution of main belt asteroids and resonances of orbit with Jupiter. In addition to the primary resonances, there are other resonances that are weaker, as evidenced by the tiny dips that are shown. The less places Jupiter interacts with the more powerful resonance.

Credit: NASA

Also, Jupiter can eject bodies from the belt and collisions could send one or both of the objects to a different orbit but light rays can be sufficient.

The Yarkovsky effect

Yarkovsky effect Yarkovsky phenomenon is small acceleration that is triggered by sunlight that is acted on asteroids. As with everything else in the Solar System, the surfaces of asteroids take in light and heat that comes from the Sun before exchanging it to be released back. But since asteroids rotate and rotating, light is reflected back at a different angle the direction from which it originated. This results in a slight acceleration due to the photons emanating by the spacecraft. Even though it is a tiny force and can last for millions or thousands in years, this force could influence the orbital path of an Asteroid. This is simply photon propulsion, an object powered by light the course of. The concept was first proposed more than 100 years ago, but it wasn't until recently that evidence of it.

Figure 5.7 Propulsion by photons, also known as "The Yarkovsky effect" in the action. Of course, this is a problem for any body that rotates and not only asteroids. Although they reflect larger amounts of light huge mass of a planet can make the impact of this force so tiny that it is minimal.

It was through the Arecibo telescope, the largest single telescope worldwide, this phenomenon was demonstrated. With a diameter of more than 300m The telescope is one of the strongest capable of pinpointing objects in the deep space. The asteroid 6489 Golevka was observed over twelve years. Based on mathematic models, scientists forecasted the location of the asteroid after this period , with or not using the Yarkovsky effect. Even though Golevka is just about a hundred meters across due to the power of the dish located at Arecibo that the location of the asteroid could be pinpointed to within just a few meters. It was amazing that in the space of 12 years it's location was discovered to be around 15km farther than it would be without Yarkovsky effect. Yarkovsky effect.

Thus, photons, the non-massive fundamental particles in light can indeed exert a small force. The asteroid, which is a only a few hundred kilometers across it has a constant force (about 1 inch - similar to an ice cube on Earth) used, which over the course of millions of years may be so powerful upon its orbit.

The path of an asteroid be affected through this effect, but the spin can be altered as well. The closer an object is to the Sun and the greater amount of light is reflected off it, and the more powerful the Yarkovsky effect is likely to be. We've already observed that NEAs are more diverse in the range of rotations than MBAs and that a large part of the reason is gravity interactions with Earth as well as Venus (and Mars to a lesser degree). Yet, the bodies which spin rapidly, and even more than times per hour, and those that move extremely slowly, are not explained by gravitational interactions. This is where the Yarkovsky effect is at play. Because objects close to the Sun are more likely to have light incident on them and the Yarkovsky effect will increase Simulations have demonstrated that asteroids that are small, smaller than 100 meters in diameter, are capable of spinning in very quick rotations (of around a few minutes) within less than a million years.

The Yarkovsky effect may also be the reason for the large proportion of NEAs that have retrograde spins. If we go back to the 'asteroid' that is shown in Figure 5.7 We can observe that it revolves around in the Sun at an angle that is anticlockwise (along with the majority of material on the disc) however, it rotates around its own axis in clockwise fashion. As you can see, the retrograde movement, accompanied by it's Yarkovsky effect, generates an energy force that operates in the opposite direction to the direction of travel which slows the speed. In the process of slowing down the asteroid, it will be pulled inwards towards the Sun. Asteroids with prograde spins are pushed outwards due to using the Yarkovsky effect. This is why the Yarkovsky effect applies a specific degree of selectiveness to asteroids. Those that have retrograde orbits are most likely to be pulled towards the inside, while those with prograde orbits are tend to get pulled outwards. Although it's only one of the forces that act on asteroids. It could explain why a greater percentage of NEAs have retrograde orbits rather than MBAs.

Figure 5.8 The dish situated at the Arecibo observatory, located in Arecibo, Puerto Rico, has a total of 1000 feet (305 meters) across, making it the biggest telescope of its kind worldwide. It is constructed into an organic hollow in the ground , and can't be controlled as an ordinary telescope. Similar to an optical telescope that is able to capture light and radio telescopes capture radio waves, however the term "radio telescope" is not entirely accurate - they're more like aerials. It is not possible to see at them and typically they are an image on graphs. This is why the asteroid 6489 Golevka was detected and the Yarkovsky effect was proved.

Credit: National Astronomy and Ionosphere Center

Light's effects on the body are not significant and if an asteroid's orbit is disturbed, and then pushed away from its normal path it could take any direction. Asteroids that have been released from the belt are placed in direction of the biggest world that is part of the Solar System, Jupiter. Jupiter's mass is greater than 300 times larger than the Earth which means that its gravitational pull can have an enormous impact on any objects that cross. In fact, Jupiter has been proven to be shields for Earth by sweeping away many massive objects that could otherwise make it into the Solar System. We have seen objects that have crossed too close to being swallowed up by Jupiter's immense size and even causing impacts larger than the Earth.

Trojan Asteroids

We've seen that the majority of asteroids are located in the belt of the main and we've just talked about the forces that regulate their orbits. However, a fascinating kind of asteroid orbits further away and has a 1:1 resonance Jupiter. These are called Trojan asteroids. They follow Jupiter's orbit, traveling in the same direction, along the exact same path as the giant planet. The first one, 588 Achilles was discovered in the year 1906. Hundreds of others are recognized, both ahead of and behind Jupiter. They are controlled by the Jupiter-Sun gravitational structure. It was the French mathematical genius Joseph-Louis Lagrange worked out the orbital mathematical equations in the 18th century and predicted the existence of Trojans for 134 years prior to the time it was discovered. He discovered there was a problem with a dual-body model, there are five places in which a small object could be in the same place relative to two larger objects. The five points are referred to by the name of "Lagrange" points. They are those points where the gravitational pull from the two big objects provide precisely the forces needed for a smaller one to revolve around them.

Figure 5.9 The position in the position of five Lagrangian points with respect to two objects with different masses In this case, that is the Sun as well as Jupiter.

3 of these points that are L1, L2 or L3 are to the two main bodies however, they are not stable; objects are easily removed if they are disturbed. The two other spots, L4 and L5, are very safe; they are the locations that, in the context of the Sun-Jupiter pair, the Trojan asteroids converge.

The Trojans aren't exactly on the Lagrangian locations, however they instead move in a gentle manner around it, demonstrating the stability of the region. Once they have been locked into the positions they are in They move in a slow manner along clear routes. The collisions between Trojan asteroids are uncommon more so than the main belt, and so their surfaces have not changed over the past few million or thousands of years. They're probably near-perfect models of the beginning of the Solar System.

Figure 5.10 The regions in which Trojan asteroids are located. The fifth and fourth areas are fairly stable, and that is the reason Trojans have settled in an extensive space.

The majority of Trojans found in the Solar System occupy resonances with Jupiter. Because of the massive gravity of Jupiter as well as its close proximity to the asteroid belt that has plenty of bodies that are candidates, the Jupiter Sun pairing has the highest concentration of Trojans in the Solar System, by far. However, there are other objects that orbit near each of the Lagrangian points of a variety of other planets, including Earth, Mars, Uranus and Neptune. Saturn has two pair of Trojan moons. Small Telesto and Calypso are located at L4 as well as L5 points, respectively (ahead as well as behind) in the more massive Tethys. Moon Dione also has two pairs, Helene as well as Polydeuces in L4 as well as L5.

Near-Earth Asteroids

There are a variety of Trojan asteroids. Trojan asteroids are examples of bodies that were released outwards of the main belt and have taken up an "follow-the-leader" strategy with Jupiter. Other bodies have longer an elliptical orbit, and are able to be close to, or even within the orbits that orbits Earth, Venus and Mercury. They are referred to as "Near-Earth Asteroids" and any one of them could strike the Earth in the future.

Astronomers constantly search for any of these stray objects tracking their orbits, and calculating their possible paths with the intention of avoiding collisions, or giving the time needed to plan for one that may be expected to happen. Since these objects are closer to the Sun and move at a much faster speed than those inside the central belt. This is why they leave long tracks on photographic or computerized images which increases the chances of being discovered.

There are three main types of NEAs The three classes are the Atens and the Apollos as well as the Amors. The two main groups are based on eccentric orbits which may traverse the Earth's orbit into space. The Atens especially are usually near to the Sun and never much further than Mars. The Amors always remain outside of the Earth's orbit and some even cross Mars.

Figure 5.11 Near-Earth objects. The routes of three common NEOs include 1221 Amor 1862 Apollo and 2062 Aten which is the name for which these groups were named. Apollos traverse the orbits of Earth as well as Mars While Atens and Amors are the only ones that cross both. Atens and Amors only cross the orbits of both Earth and Mars and Mars, respectively.

There are over 10,000 NEAs that are known. However, there is one kind which is crucial to find ones that exceed 1 km in diameter. It is the size that we're considering catastrophic global destruction in the event of an impact. There are more than 90 such objects on Near-Earth orbits, but fortunately they are not expected to strike the Earth in the next century. However, objects smaller than 1 km in size could cause significant harm There is a multitude of them, or even millions that are hundreds of metres wide, and some extremely dangerous objects are not identified. The majority of NEOs do not come close to Earth However, as they are orbiting in regular intervals certain come close frequently.

Potentially dangerous objects

PHAs are asteroids that are extremely close to Earth. Particularly, all asteroids which fall within 0.05 of an AU (4.65 million km , which is roughly 20 distances between Earth and Moon) and are greater than 150 meters (500ft) within diameter are considered PHAs. There are currently around 1400 PHAs that are known to exist. The most important thing is that being a PHA implies that it has the potential to be struck by the Earth but it's not guaranteed the likelihood of it happening. By keeping track of and adjusting the orbits, we can reduce the risk of impact, and be prepared for an impact that could occur.

Figure 5.12 The differences among the orbits the typical close-Earth-sized asteroid (blue) as well as a dangerous asteroid, known as the PHA (red).

Close-up views

A number of spacecrafts have passed by and observed, and even landed on asteroids. The Galileo spacecraft passed by nine asteroids, 951 Gaspra in 1991, as well as the 243 Ida in 1993. NEAR-Shoemaker looked at the asteroids 25,3 Mathilde as well as 433 Eros as well as Deep Space 1 and Stardust each had close encounters asteroids.

Figure 5.13 The size of the object matters A collage of comets and asteroids that were that were photographed and visited through space craft (up to 2010.). Lutetia was, prior to the time the Dawn craft came across the 4 Vesta in 2011, the biggest asteroids photographed in the air (by Rosetta). Its composition is not clear and it is classified as an M-type asteroid, however it is believed to be of the characteristics of a carbonaceous one.

Credit: ESA, NASA, JAXA, RAS, JHUAPL, UMD, OSIRIS;

Montage Emily Lakdawalla (Planetary Society) and Ted Stryk

Hayabusa examines a pile of rubble

On May 9th 2003 The Japan Aerospace Exploration Agency (JAXA) launched the MUSES the C spacecraft into the body near Earth 25143 Itokawa. After its successful launch, the spacecraft was named Hayabusa which is Japanese meaning "falcon". It was powered by an ion-drive motor, Hayabusa rendezvoused with its destination on September 30th in 2005. It then settled on a location that was about 20 kilometers from the asteroid, which was moving to the same speed as its target while both orbited the Sun.

Hayabusa was able to make a landing if it wanted to, however it had a small craft called Minerva, (about half a kilogram and 10 centimeters across) which was specially made for landing on and looking at the Asteroid. The falcon was able to descend in a descending manner from 1.4 km at 3 cm/sec, to 55m, from where it was scheduled to launch Minerva before rising again. But at the time Minerva was ready to launch the craft's parent had already reached the 55m mark and had begun its ascent. However, the launch came too late and release came at a higher altitude than originally planned. The contact failed with Minerva and the insignificant craft sank into space, never to be ever seen again.

On the 19th November the falcon itself landed on the area, and stayed on the surface for about 30 minutes before taking off. The following week, on the 25th of November, it briefly slowed down, before rising up again in the air, as if a bird is hovering to look for the fish in the ocean. After collecting some samples it set off toward Earth returning home in June 2010. Hayabusa was the only craft to be able to land on an asteroid, as well as the first one to collect samples and return them home.

Figure 5.14 25143 Itokawa captured from the Japanese Hayabusa spacecraft in 2005, is 500 meters long. Instead of a single, solid rock, the asteroid appears more like the shape of a pile of rubble that is loosely tied together. The difference in smooth and rough terrain is likely caused by the impact that rubs the asteroid's surface, which separates the large and small rocks.

Courtesy of JAXA

Itokawa is itself a small rock that is only 500m in diameter, which rotates every day twice. Its orbit for one-and-a-half years puts it within the Earth's orbit around the Sun and also close to the orbit of Mars. Accessible, it's an ideal subject for research of small objects that are moving close to and could meet with Earth. An oblong, rugged form, it is composed of an incredibly tiny "head" as well as a massive "body" and an elongated "neck" connecting the three. The body and head could be separated and then slowly became one by rubbing them together at a steady speed. The neck could also have been pushed back by contact with a larger or even a large.

A heap of pieces and bits that are loosely bound by gravity. the absence of visible impact craters isn't caused by the absence of collisions but likely because they are filled by the rubble after the asteroid is exposed to the vibrations and impacts. Itokawa's vital data show its fluid nature, with 3.58 1018 kg, with a weight of around 1950 kg/m3, which is significantly lower than terrestrial rocks (see Section 4.). This suggests a porous interior with air gaps, many of which are filled with sand. Most likely, Itokawa as well as other loose bodies, were created through the break-up early of a parent, followed by the recombination of pieces.

Chapter 8: The Moment Worlds Collide...

A comet strikes Jupiter

On a cloudy, dark evening in the month of March in 1993 Eugene Shoemaker, his wife Carolyn as well as amateurastronomer David Levy were continuing routine observations of comets and asteroids which might be heading for Earth. Utilizing a telescope in the Palomar Observatory in California, Carolyn noticed a large "squashed comet" when she was looking through some of the pictures. Higher-powered telescopes were used and, with higher resolution, confirmed that the stretched out blur to be a few tiny comets that were aligned across the sky. Because it was (or in fact, the were) the ninth object of short-period observed by the three of telescopes, the fragments were identified as"Comet" "Shoemaker-Levy 9". Through analyzing its path, scientists realized that it had come extremely close to Jupiter in the year prior; the earlier single body could have been traveling around Jupiter for a long time but when it got too close to Jupiter, it was pulled apart due to the differences between the far and near-side gravitational pull. It was smashed into at least 20 visible pieces, these instances were not unusual but this was a first because it was bound for the collision with Jupiter. The date of its fate was estimated as July 1994, which was two years after the event and one year following the discovery.

Numerous craters are seen on the bodies of many bodies in the Solar System. There are numerous craters on the Solar System. Moon, Mars, and the Jupiter satellites, especially Callisto are a few of the most prominent instances, however, never before has such an impact been seen. It was therefore an amazing chance to experience it for the first time in history.

The July 1994 collision was perhaps the most observed moment throughout the entire history of Astronomy. In the week between the 16-24th of July in 1994 virtually all telescopes were pointed towards Jupiter. On Mauna Kea in Hawaii, the 10-meter telescope of Keck Observatory Keck Observatory was ready to capture infrared wavelengths while Hubble Space Telescope Hubble Space Telescope (HST) was observing the events in visible wavelengths. The collisions took place on the other part of Jupiter and were obscured from the view of observers. But, Jupiter rotates every 9 hours and 55.5

minutes, which means that the sites of the collisions were swiftly revealed. In the Galileo spacecraft, on its way to Jupiter had a clear observation of collisions.

Figure 6.1: The last journey of the Shoemaker-Levy 9. After orbiting for over 50 years. Its spacecraft was destroyed after it crossed only 0.0006 (AU) (90,000 km) from the center of Jupiter. Because Jupiter has a 70,000 km its radius and the body was 20km from the peaks of the cloud. The fragments traveled up to 0.3 Aura away (just when they were first found) before being pulled back.

When the fragments collided, they was exploding with the force that hundreds of millions of bombs from nuclear war. Gas plumes that erupted hotly upwards were observed by Galileo as high as 20000 Kelvin in temperatures.

As the train carriages which stalled and one after another, the pieces crashed into Jupiter each leaving an indentation larger than the Earth. The entire planet seemed shake from the impact. powerful winds resonated throughout the planet. Over the course of time the marks faded away as they were absorbed by the air.

Images taken by images taken by the Hubble Space Telescope (HST) were captured during January of 1994. around one year after discovery and six months prior to the collisions. The pieces were scattered across 1.1 million kilometers, which is more than three times that distance Earth as well as the Moon. The largest piece measured about 2 km wide.

Figure 6.2 The remains of the comet Shoemaker-Levy 9 in July of 1992 the comet travelled close enough to Jupiter that the icy material within its nucleus was torn apart due to the gravitational forces of the planet. This is a compilation of multiple images taken with HST Hubble Space Telescope (HST) in January 1994, just six months before they fell into Jupiter. The piece of string measures about 1.1 million km in length, and the largest is approximately 2 km in length.

Credit: NASA, STScl

Figure 6.3 Impact of comet Shoemaker Levy 9. The Hubble Space Telescope image shows various darker areas (top) in Jupiter with each one marking the location of impact for a piece that came from the spacecraft. The Earth-sized marks remained approximately five months, before they were pulled apart by the winds of the atmosphere's outer layer.

Credit: H. Hammel (MIT) H. Hammel (MIT), HST, WFPC2, NASA

The Sun devours...

Many comets are believed to have sunk into the Sun at times, touching the Sun's the outer layer of its the atmosphere (the corona). Many of them have been photographed with the Solar and Heliospheric Observatory (SOHO founded in 1995) which was developed with the help of NASA as well as ESA to examine the Sun in depth starting from its center to the solar wind. One of the instruments onboard (the Large Angle and Spectrometric Coronagraph LASCO) LASCO) utilizes an occulting disc that blocks off direct sunlight from the primary part of the Sun and creates artificially obscuring the Sun's light all day long all week long. This allows the detection of comets as they pass into the sun's corona. Sometimes , when pressure is applied to them, they dissolve while other comets are completely destroyed long before they get to the same part that is the Sun. Most comets that hit on the Sun belong to the "Kreutz sungrazer" group. It is believed that they are fragments of a larger comet which broke up several thousand years in the past when it was closer in proximity to the Sun. They were named for their 19th Century astronomer Heinrich Kreutz who realised that they had an identical origin and follow similar orbits. In spite of their spectacular display they are incredibly tiny, measuring between 6 and 12 meters across.

Amateur astronomers worldwide have examined the real-time images from SOHO posted at http://sohowww.nascom.nasa.gov/, and discovered hundreds of new comets, including some sungrazers.

Photo courtesy of SOHO LASCO consortium. SOHO is a joint project of international collaboration that is a collaboration between ESA along with NASA.

Impacts and Collisions

If we look at Earth's history that spans millions of years it becomes obvious that it is not surprising that impact of asteroid is part of the Earth's existence. In addition collisions and impacts have been, and continue to be crucial to the development and formation of our Solar System.

On one side of the scale , we find low collisions that have low energy. They were the reason for the creation of the Solar System in the first place. Micrometre-sized particles of dust rub against one another and then merged and eventually formed larger bodies such as protoplanets, asteroids and finally planets similar to those on which we are based.

At the opposite end on the spectrum high-energy collisions created the craters are seen on many objects nowadays. While major collisions were frequent in the beginning of the Solar System, they are extremely rare today. Sure, we wouldn't like to see one happening on Earth However, we might are a part of these events as they were (and remain) essential in the transport of every kind of material across the Solar System body to another. This includes organic matter that could be instrumental in kicking-starting our existence on Earth as a concept referred to as panspermia.

A number of planets were drastically changed by collisions that occurred early in their life. Venus as well as Uranus exhibit significant tilts in their axes, believed to result from significant collisions with other bodies of similar size. The Earth has an inclination of 23.5o towards the plane of its orbit, but Uranus is rotated in a lateral direction, having the tilt of 97o and Venus turns upside down at 177o, whirling around the other direction to that it revolves in an orbit that is retrograde. The Earth also suffered a massive collision in the beginning of its history: the Giant Impact theory dictates that the Mars-sized object struck the Earth and ejected a massive chunk of material that became our Moon.

Shock Features on Earth

The most obvious sign for an accident is an impact crash crater. There are plenty of excellent examples of this in the Solar System; just look at the Moon in the night sky. Each of the terrestrial solid planets, along with numerous moons of gas giants, are an account of our Solar System; the scars of the ancient collisions with space-speeding objects.

The worlds that aren't geologically active, such as Mars are among the most easy planets to search for the craters. Their sizes an indication of the heavy bombardment that these worlds have experienced.

Figure 6.5 The surface of Callisto is the second largest moon of Jupiter's and the third largest of the Solar System. The craters are covered with craters. The bulls-eye in the left is believed to be an impact basin that was thought to be been formed at the time that Callisto was a young moon. The crater measures 600 km wide and the entire basin extends for 2000 kilometers. This is around as big as Mercury (4,800 kilometers across). The image was captured from NASA's Voyager 1 craft from a distance of 350,000 kilometers.

Credit: NASA

Finding craters may be simple. Large bowl-shaped dents are apparent. On Earth prominent craters are usually young. Those which are millions, or million years old are usually buried beneath rocks, sand, or water, covered by vegetation, or destroyed or completely erased. The shape and form of a landscape is typically not sufficient to show the existence of an apex. However aerial photography has revealed several rounded areas that upon closer examination are revealed to be craters or at the very least, fragments left by one.

Figure 6.6 Figure 6.6: The Clearwater Lakes in Canada, as seen from the Space Shuttle. Around 290 million years old the lakes were formed through a pair of binary asteroids that struck each other and the Earth together. The lakes are not distinct and are one huge lake that covers two distinct depressions. The largest, located to North West, is North West, is about 36 km wide, with islands in its center. The smaller measures 26 km in length and an island line divide the two.

Credit: NASA

Rock that has been struck often keeps the evidence of the shock very well. In some cases, small diamonds may be seen, created by the massive pressures generated through the collision. Common are also glass beads that have been melted as well as minerals such as stishovite and coesite. They form when pressures of between 10,000 and 100,000 times the force from the air are applied on quartz.

It is true that the Earth has been impacted by the elements throughout its time, but the processes of geology like volcanism and weathering have obscured or erased a lot of craters. Perhaps 180 excellent examples remain scattered all over the earth. The biggest craters are at Vredefort within South Africa (300 km across) and Sudbury in Canada (250km) are believed to be among the oldest, with around 2.25 billion or 1.85 billion years, respectively.

We are aware that the impact rate in the present is significantly lower in comparison to the millions of years earlier. the only explanation that can be made would be the fact that there's less objects floating about in the Solar System. The majority of asteroids are in orbits around Mars and Jupiter in which they do not pose a threat to our existence, however certain ones have been pushed to leave these orbits and have taken more eccentric routes, frequently getting close to, or even traversing the earth's orbit. Earth.

Impact Rates

An easy way to estimate the impact rate with the Earth is to look at the age and number of craters found on the Moon. Because the Moon is in close proximity to the Earth so it's sensible to suppose that the cratering rates for the two objects will be equal. It is also useful to note that the Moon is a geologically inactive object i.e. there aren't any volcanoes or earthquakes and no atmosphere, therefore there is nothing to stop collisions, or slows down any rocks that come in. This means that when craters are formed, they remain relatively unchanged, except for the impact of other objects at the same time (large impact craters become damaged over time due to the impact of smaller objects that sit on the top). If we study the size, the number, and age of craters found on the Moon and determining their age, we can determine the impact rates for the Earth.

The craters that are visible on the Moon as well as on Mercury and Mars as well as other relatively unaltered bodies - show a time of intense bombardment of the Solar System some 4 billion years ago. As time passed, the amount of debris within the Solar System decreased - pieces were taken up by the wind and then washed away, or removed away from this Solar System altogether. The rate of cratering today is significantly lower than it was in the past and a more detailed review of the hazards of impacts and rates is provided in the Section 9.

Explosions from the sky

Even with very little coverage, we are aware that space invaders pound into our globe frequently. It is believed that the US Department for Defense has been tracking these objects for years and military satellites, who keep the watch for hostile rockets and nuclear explosions have observed the explosions that occur when large, house-sized rocks fall off the high-earth. When they are heated to incandescence the boulders break up and then vanish completely upon the floor. Defense network on the ground has verified these reports and, in actuality there's an explosion the size of a nuclear bomb that is occurring each month in the atmosphere! These studies could help to maintain the peace of the world. However, mistaking an natural explosion as a man-made nuclear attack could be very dangerous indeed!

Set collision course Then go full speed ahead! While Jupiter protects us from any rocks that could enter the inside of the Solar System, it can deflect many of them out of their orbits. The Chelyabinsk meteor may have been ejected off the Main-Belt by oneof, or a combination collisions as well as the Yarkovsky effect and the gravity of Jupiter will have all played a role in. Through observing its location and its descent through different camera angles, and effects like the shadows that it created, we can determine the exact path it traveled through the atmosphere, as well as the exact position at which it reached. By doing this, we can trace its route back to deep space. The meteor was able to complete a 2 year orbit, which was at an angle with the Earth. It was circling its orbit for thousands, hundreds of thousands, possibly millions of years, and then the day it spotted another planet that was in the way.

We are aware of where asteroids originate and the forces that influence their lives. What is the next time? How do you react when meteorite hits? (This fact isn't that difficult to think of!)

When the meteor struck our planet it was traveling at around 60,000 km/h (18 kilometers/sec). Although it was a whopping 20 metres in diameter, and the weight of 7,000 tons but only a small amount of the rock was recovered. The immediate effects were felt across an area of 3000 square km How could such a small thing cause such destruction?

An object of this size struck was over 100 years ago. studying the consequences of this event, we can better understand the causes of such incidents.

On June 30 1908, a massive explosion destroyed the forest in Tunguska, Siberia. The site was remote, and it took more than two decades did Russians make an attempt to reach the area. When they finally did, they found amazing facts: over an area that was the size of London 16 million trees were not just fallen and fell down, but completely crushed. A meteorite impact was thought to be the reason however, how do you determine the lack of a crater of impact? Similar to that of the Chelyabinsk meteor, this destruction at Tunguska was likely to be the result of an impact of a particular type.

A side note: There might be investigations into the site prior to being published, but major the events of World War 1, the Russian Revolution and the Russian Civil War were prioritised at the period.

Tunguska Impact

Figure 6.7 What happened after The aftermath of the Tunguska Impact. The 30th of June 1908, a massive fireball was spotted across the sky, above the Tunguska River located in Siberia, Russia. More intense than the Sun the fireball burst through the atmosphere, destroying the forests for miles across all directions. The trees were able to show that the cause of the blast to excavators when this photograph was taken nearly two years after the fact.

Credit: Leonid Kulik Expedition, Wikimedia Commons

We now know that the blast at Tunguska was the result of an object which entered the atmosphere and came near the earth's surface but then burst into flames before hitting the ground. The winds of hurricanes which flattened thousands of square miles are referred to as an air blast which is an explosion that occurs prior to directly colliding to the earth. The largest meteorites are the only ones that reach the earth; the majority break up into the air. The heat produced by a blast that pushes through miles of air every second, forming a single piece is difficult. But, the powerful winds produced by the air blast are strong in their own right; the ones which leveled Tunguska were caused by an object as tiny as 40m in size.

The Chelyabinsk impactor was also less than 20m in size and was only 1/10th as powerful of the predecessor. Due to its tiny size it was unable to reach as deep into the air before breaking upand breaking up more than twice as far away from the ground. This resulted, thankfully with less destruction to the ground. However, if we were to be hit by an object that is bigger than that at Tunguska It could remain on the ground, similar to the object that caused this:

Barringer Crater

Figure 6.8 The figure is the Barringer Crater in Arizona, USA. It is one of the spectacular impact craters found on Earth The Barringer crater measures 1.2 km in diameter and 175 meters deep. It is named after Daniel Barringer, who first suggested it could be the result of a meteorite collision the massive hole on the earth's surface was carved out 50 years ago. Similar impacts today could cause the destruction of any city however the rock that caused it was only 50 meters across! What's the distinction from the Tunguska site and the Barringer sites? Simplyput, Barringer was the site where the Barringer crater was destroyed directly by meteorite, while the destruction that occurred at Tunguska was the result of an explosion above ground.

Credit: D. Roddy (LPI)

Fortunately, we don't need to wait for another massive rock to strike the ground to discover why they're so damaging We can even simulate impact in laboratory.

The Vertical Gun range of NASA

NASA's Ames Vertical Gun Range (AMGR) was developed in the Apollo space flight period. Its objective was to discover the way that the moon's surface would react to impacts from landings of spacecraft. Nowadays, it is utilized to study impacts of asteroid.

With its light-gas and the powder guns The AVGR can fire projectiles at speeds ranging from 0.5 up to seven km/s. Its angle in relation to the target may be altered from a head-on collision , to only the glancing of a strike. This is essential to determine the way that the angle of impact influences the form of crater formed.

There are a variety of shapes available that range from regular spheres to jagged objects, and even clusters of tiny particles are able to be fired. In addition, the projectiles they are metallic or mineral-based (e.g. quartz) or glass and range from one-quarter of an inch up to 1/8 inch (marble size).

The target chamber measures approximately 2.5 meters in height and the diameter. It is able to maintain an atmosphere that is less than the thickness of 1/25000th of the Earth's atmosphere and is able to be filled with different gases to mimic the atmospheres of the planets.

Sand is a good targets since it captures the impact very efficiently. Outside the chamber, cameras with high speeds that can film up to 1 million frames per second, capture every aspect of the impact as well as the consequences.

Through simulations like that, one can determine that in a realistic scenario the entire region of impact will be smoldering. Even outside of this area there'd be no satisfaction; the winds would be strong enough to raise houses in the area and drop them in the distance of hundreds of miles! Such experiments could uncover a variety of important aspects:

First of all, it's not just the impact that causes the damage, but also the smoke that is emitted from the length of the range. It is possible to observe the events in real-time and observe the process taking place however, an asteroid strike causes more destruction than is implied by the impact itself. Comparing the effects of an air explosion with the effects of a ground strike suggest that Chelyabinsk did not suffer any serious damage. It is believed that the biggest piece of debris to strike on the floor at Chelyabinsk was only 500kg in weight, which is a tiny only a fraction of 7000 tons.

Ground strikes are the most destructive natural danger known. From space Earth's encounters with massive asteroids during its lengthy time are seen. There is evidence from the past of a catastrophic impact which changed the nature of life on Earth.

Impact Craters

In the history of mankind there haven't been large craters that have formed. While we are able to simulate the effects of impacts in labs but we are unable to replicate the creation of a massive impact crater. Therefore our understanding of these abrupt massive energy releases along with the resultant crater is only determined by what we see through research both scientific and theoretic and also the analysis of geological features of massive impact craters that have been discovered on Earth.

The concept is basic and simple, and is it has been seen in movies that show a large rock hitting the ground, leading to massive destruction and even death! The process of creating a crater is incredibly complicated There are many details that remain unanswered.

The Shockwaves as well as the Speed of Sound

The development of craters from impact is controlled by various variables. The dimensions and speed of the object that is hitting are probably the most evident. To be able to withstand the descent into air without harm, the object needs to be big - it must have a diameter that is greater than 50 for a solid object or perhaps 20m for one composed from metal (iron) and moving at speeds of up to 10km/s (36,000 kilometers/hour).

The projectile that erupted into the Barringer crater was around 50m in size, which is the minimum size to reach to the earth in one chunk. Anything smaller than this will not make it to the ground in a single piece Although a lot of fragments are scattered over the ground, the majority of them leave little more than a tiny mark upon landing. The biggest fragments physically break up soil, but they lack the energy required to produce the shockwaves required to create large craters.

Fragments that impact the ground are not really of interest We will focus us with high-speed impact craters, which are created by an object that is mostly intact and with no noticeable acceleration in the course of its descent. The first step is to know the definition of a shockwave.

Question If an unintentional tree falls in the forest but no one is around is it making an sound?

Certain people will respond "No", "sound is subjective and requires the attention of a person." "Yes", others will say, "A sound is an objective thing, not something that is inside a person's mind". Of course , arguments about this kind of thing are not worth the effort since people aren't arguing over the definition of what sound is however, the definition of what it means. The answer will depend on the definition we use to define sound. A physicist could be more objective, and state the sound we hear is simply a type of energy that is present regardless of whether it is heard. analyze the nature of sound.

The majority of sounds we hear in our daily lives are transmitted by air, however any substance that is able to change its shape or form whether liquid, solid or gas - is able to transmit sound. The air, as well as gases generally, are not good for sound transmission. Since the molecules within gases are in essence, separate from each other, they cannot vibrate against one another very effectively. They need to move or collide and transfer energy in this manner.

Liquids and particularly solids contain molecules that are physically connected, and so can interact without needing to travel far.

The test is simple by placing your device on the table a far away, it could be too quiet for you to hear, but just put your ear near the table then the audio will come through evident. Also, by putting your ear against the ground, you will be able to hear sounds made at a distance.

Figure 6.12 The speed of sound in both rock and air. When you're in air (top) the molecules are dispersed, making it difficult to transfer energies from one the other by contact. When you are in the case of solids (bottom) they have the molecules joined therefore transfer of energy via vibration is easy.

Imagine this on a much larger scale, using marbles as the molecules. If we have a line of marbles that are in contact the moment the one is fired in the one end, it is pushed out of the other end nearly immediately. If the marbles are separated and are not in contact, it takes a longer time for one of them to be released from the opposite side.

In a vacuum, there's no molecules, which means that sound can't travel because there isn't any medium to transfer the energy. In the air the sound waves travel at around 335 milliseconds. On the other hand, in rock, sounds travels between 5 and around 8 or 9 km/s based on the particular nature of the rock.

However, since an impactor typically strikes the rock with speeds in excess of 10 km/s, the force isn't able to travel through the rock at the speed it needs to. Due to this, the energy quickly builds up behind the front of the wave and an extremely high-pressure shockwave is formed. The sonic boom that occurs in the air is an illustration of a shockwave created when an object is at that speed at which sound travels in the air.

The pressure generated through an impact could reach hundreds of GPa (Giga Pascal). To put this in perspective it is that there is a pressure that hovers over the Earth generated by the weight of atmospheric pressure (atmospheric pressure) is around 100 kPa. Rock is able to resist at least 1 GPa which is 10,000 times the amount of atmospheric pressure. But over this point it expands and deforms like liquid. So, when it is struck at an extremely high velocity the surface of the rock is distorted and pushed upwards by the shock waves, thus creating the crater.

The development of an impact Crater

Contact between the projectile and its target and digging the hole and the immediate modification of it are simple, but complicated processes. Somewhat arbitrarily, it's ideal to divide things into three distinct phases: Contact and Compression Exploration, Modification, and Contact.

1. Contact and Compression stage

The phase begins at the moment when the front end of the projectile comes into contact with the ground. To simplify things it is assumed in all instances that the target is a solid rock and not sand, water or any other type of material.

For just a short period of time it continues moving as if nothing has taken place. Thanks to the power, the stone is compressed, and then is pushed away almost without resistance. After initially penetrating to its own diameter in the target and then overcoming resistance to its movement, suddenly ceases to move.

In this stage, the shockwaves propagate outwards from the contact between the target and the collision which carry the bulk of the projectile's enormous energy kinetics. It is important to note that this is below the ground's surface since the impacter is already infiltrating deep into ground. Similar to the target, the projectile itself gets compressed when it comes to a stop. Because some of the energy kinetically generated transforms into energy, the heat created is so high that any the solid material, like rock, can begin to flow like liquid.

Figure 6.13 The Compression and Contact Phase. A cross-sectional view of the moment when a massive cylindrical object strikes the target in a uniform manner (top) with just a fraction of seconds after the impact (bottom). It has penetrated approximately half of its length into the target. The shockwaves radiate outwards out of the interface with the object and.

The shockwaves travel all around including upwards and back to the projectile. Consider first the shockwaves that are traveling down: they expand outwards , breaking up and compressing the rock through which they travel. However, they gradually diminish in intensity as they expand to an even larger size. In addition the energy used to heat and form the rock.

When the projectile is struck the pressure may exceed 100 GPa (1 million times atmospheric pressure) which can cause melting, or even burning, the projectile, and the surrounding rock. Pressures in the 10s of GPa can be present for a long distance away from where the impact occurred, dramatically altering the shape of rock that is not melted. Further away, when the pressure is reduced to 1 GPa and the shock waves transform into normal seismic waves and their speed decreases to the normal sound speed inside rocks (5-8 kilometers per second). These waves are able to travel across the entire Earth as similar waves generated by volcanoes and earthquakes. They don't cause any permanent damage to the rock they pass through but they are strong enough to trigger cracks and fault lines and landslides close to the surface. Once the pressure is to below the limit of resistance for this rock type, then the edges of the crater will eventually take shape. Let's now look at the shockwaves which travel upwards through the projectile and out to the side towards the surface of the earth. They travel in a normal manner until they reach the back of the projectile or the surface of the ground after which they are reflected back down again. The reflected waves have an entirely different shape from the initial shockwave. While the original shockwave compressed the material that it was moved by, this reflected one is the reverse and pulls the material apart which is known as rarefaction. As the rarefaction wave passes through the projectile and breaks the projectile and then releases (or releases) it outwards and upwards. The particles released upwards swiftly disappear. When the wave is at its front and the projectile is completely destroyed then the contact and compression stage is considered to have ended.

The projectile has no effect on the impact. the excavation portion of the impact crater completed by the release wave and the shockwave spreading outwards to the targeted area. A portion of the vapourized rock escapes as a part of the vapour plume, while the remainder is blended with the target.

The duration of the Contact and Compression phase is based on the size and speed of the object being impacted and should not exceed just a few seconds for the biggest objects. The time it takes for an object to break their own dimensions into target is the same as the diameter of the object divided by the speed. Therefore, a 5-kilometer wide rock traveling at 10 km/s is only one-half second. The "squashing" duration for the projectile is just approximately 5 times that depending on the properties of the projectile as well as the goal. Many projectiles are smaller than 5 km wide, so the majority of contact and compression phases can be completed in one second.

2. Excavation Stage:

This phase begins right after the contact or compression phase. Here the impact crater gets made open by the intricate interaction between shockwave as well as a rarefaction waves that radiate outwards towards the earth. The shockwave is able to push material forward and compresses it, with the release wave is following is pulling the material apart before expels it backwards. The material that is ejected from the newly formed crater is called ejecta. It can be released in a matter of a second. The material then falls back into the ground, covering the crater as well as the surrounding area. The material to the sides is pulled outwards, then folded upwards, creating a rim around the crater.

The rapid movements result in a bowl-shaped depression, or a temporary crater within the direction of the target. The final size of the crater can be 20 or more times the size of the initial impacter.

Through the entire ground, the rock remains deformed and displaced if the force that pushes it is powerful enough. The shockwave eventually ceases to be strong enough to break any further material. It is at this point the expansion of the crater's transient form is finished and the excavation phase also ends.

The second stage, while more time-consuming than the previous one however, is still quite brief it could take a few seconds or even minutes for the largest craters. In the excavation process of a one km crater, like the Barringer crater could have taken only two seconds, whereas 200 km craters that only two of them on Earth over this size - will take around 90 minutes. The depth at which an crater ranges from 1/3 to 1/2 of its size.

So far, we have believed that once one phase of formation is completed and the next stage begins immediately. However, the whole length of the crater is usually not removed until the maximum depth is attained. Thus, if excavations are still taking place within the outer regions of the crater, its base might have been completely cleared and the next step which is the modification of the crater's shape - is already underway. However, the idea of a transient crater essential and all structures that impact are believed to traverse this phase - crucial for comparing craters with an entirely different size or different locations, including at the Moon or other planets.

3. Modification Stage:

After the transient crater been filled to its capacity the excavation stage is over and the alteration stage commences immediately. Seismic waves continue expanding outwards, but are not of any interest in this case.

The primary factors in the formation of the crater during this stage include gravity, as well as rock mechanics. The initial part of the modification stage is just a bit longer than the excavation phase which is a matter of a few minutes for larger structures, just under sixty seconds for a tiny one. The whole stage of modification is not a set timeframe as the collapse of the walls of the crater and an uplift central to the structure (as discussed in the following section) are followed by longer-term geological processes , such as sedimentation and erosion.

(1) Contact, and Compression stage that begins with the initial expansion of the projectile and the release outwards of shock waves. A wave that initially acts backwards is then reflected forwards to form an release wave. The the projectile is destroyed completely. (2) The excavation stage is when shock waves expand into the target, and the interaction between the release wave and ground releases material, opening the crater. (3) The release and shock waves expand continuously and a constant flow (ejecta) of materials (ejecta) releases, allowing the crater. (4) the end of the excavation phase The growing shockwave is losing a lot of its strength and not penetrate the rock anymore and the crater that was formed transiently reaches its maximum size as the wave of ejecta is stopped and an uplifted edge forms within the area of the crater. (5) Stage of modification: the crater's steep walls fall down and, with ejecta falling back, form an unmixed layer of substance within the cavity (breccia lens). (6) the final part of the crater following modification. The processes (1) (2) - (4) require a few seconds to complete, contingent on the speed and size of the projectile being used. Final crater (5) and (6)

will be formed in the course of minutes or hours. Other changes, like filling and erosion, happen on geological timescales ranging from thousands up to trillions of years. The crater that is shown here is one of the "simple" one, and is different from other, more complex craters that will be described in the following sections.

Different types of craters

Let's now look at the extent to which a temporary the crater can be altered in alteration. This is contingent on a range of elements: the primary factor is size. the crater however the physical properties of the rock and the environment around it, also are important factors to consider.

The primary factor is the size of the crater: smaller craters might be modified only from their initial transient form or perhaps just the walls of the upper crater are collapsed. Larger craters could experience significant structural changes, for example, upwards movement of the floor or the outer rim breaking under the weight of its own.

There are three main types of craters that may form: simple, complicated and multi-ringed basins.

Simple Craters

The smallest craters are bowl-shaped depressions only one kilometre across. The original dimensions are unchanged except for the fall of the steep walls, which were not supported enough. The material that was ejected falls back to the bottom of the crater which reduces its depth by nearly half. The filling material is known as Breccia Lens, and is a mixture of rock as well as other types as well as the melted rock.

The lens can be destroyed over time, or covered by a level of general sediment. The Barringer crater, which was mentioned earlier, is a great example of a crater that is simple.

Complex Craters

Larger craters typically exhibit more intricate structures, with a central peak protruding from the floor, and significant inward collapse of the rim. Central uplifts are an inevitable result of highly stretched material that is trying to get back to a state of equilibrium.

On the Earth the line between craters that are simple and complex can be found from 2 to 4 kilometers depending on the precise nature of the object. These numbers are only applicable to the Earth and the Moon. Moon as well as other planets on the terrestrial side (and the moons of gigantic planets) all are less gravity-dependent than Earth which alters the mechanisms of the formation of craters.

It is also evident how the continuous bombardment from the Moon alters its surface. towards the top right, we can see an older crater that Theophilus is laying and to the left the crater is younger and has been formed inside Theophilus itself. The spiral-shaped rod in the left-hand side is an instrument aboard that of the Apollo 16 spacecraft from which this photograph was taken.

Multi-ring basins

The biggest impact craters of the Solar System are thousands of kilometers in size. Viewed in space from a distance, the craters appear like huge bulls-eyes made of numerous concentric rings, with valleys between. These are the result of massive impacting projectiles that range from tens to hundreds of kilometers in size, they originate predominantly from the early days in the Solar System - from about 3.9 billion years ago or earlier in the time when massive objects were commonplace throughout this region of the Solar System, and therefore collisions were not uncommon. The most impressive craters of this kind aren't found on Earth however, they are found elsewhere on terrestrial objects. for instance, the Moon and Mars contain plenty however Mercury and moons of Jupiter contain some. One example is the Valhalla Krater located on Saturn's moon Callisto (see the figure 6.5). Although the actual crater measures just 600 km in size the rings extend to as far as 2000 km away from the center (so approximately 4000 km in circumference).

There are a few huge craters within the Solar System that do not have a multi-ring structure possibly due to erosion through time. One good example is the South Pole-Aitkin basin on the Moon with an area of 2500 kilometers. It is important to note in this moment that craters are formed in different ways on various bodies. On the Moon multi-ring basins are formed when the crater is more than 500 km in size. But the Moon is just 1/6th the gravitational force of Earth. Transitions from one kind of crater depending on the gravitational strength. This suggests that multi-ring-shaped craters in Earth appear at diameters that exceed 100 kilometers. A few structures on Earth are this large and most of them have been submerged or degraded. So the identification of multi-ring basins on Earth is difficult, even though the craters with the most massive size, Vredefort (South Africa 300km); Sudbury (Canada, 250 km); Chicxulub (Mexico, 180 kilometers) are likely possibilities. Two of them are approximately 2 billion years old and the third one is nearly half submerged however, with a age of 65 million it's still relatively young and seems to be in good condition (we'll learn more details about this particular crater in the future).

Most complex of all craters, the significant changes to the multi-ring crater will be clearer than in simpler craters. It is unclear if the change from simple to more complicated craters is dependent on the size of the crater, or if large basins can are only formed when there are certain conditions within the target rock, such as for instance , a thin layer of material. In addition, there are many large craters that extend over 1000 km that show clear rings, whereas others don't, something which isn't fully understood.

Chapter 9: "Death From Above" Close Encounters With Mass Extinctions Mass Extinctions

It's the Earth in Motion

As with its neighbors in the middle of the Solar System, the Earth has grown through multiple collisions in its early days, around 4.6 until 4.5 billion years in the past. Massive bodies smashing together created the Earth to its present size One of these generated enough debris that it merged into our sole natural satellite that is the Moon. A different object (and there may be a variety of objects that contributed) caused the planet to be tilted and caused it to tilt its axial at 23.5o in relation to the orbital plane. The shear energy created by asteroids constantly pounding the surface held planet Earth in a state of molten for around 800 million years. However, when approximately 3.8 billion years after the event, the frequency of bombardment decreased and the Earth began to cool.

The first, and most essential life forms, were discovered around 3.5 billion years ago. However, just a few billion years later was the planet sufficiently stable to allow complex multi-celled organisms to live. In the years that followed and for the rest of time every now and then, a meteor was able to smack and kill a variety of species. The most recent one occurred 65 million years ago. However, life has the opportunity to grow, or even flourish during the time during the time between large impact.

Mass extinction

The demise of a single species is an essential and inevitable aspect of evolution. However, based on fossil records , we can observe that in some instances, a number of species end up dying all at once. Although we aren't sure about the cause of each of them through the combination of geologic and fossil records it is possible to speculate on what is behind the majority of these. The largest mass extinctions in records include the Permian Triassic extinction that occurred 251 million years ago. This wiped out more than 90 percent of marine species as well as 70% of species that lived on land. The destruction of this magnitude needs an enormous amount of energy to be released and either an collision with an asteroid or a massive volcano eruption could be the probable causes. It is believed that there have occurred five massive extinctions in the last 500 million years.

The more recent Cretaceous-Tertiary era that occurred 65 million years ago may be the most well-known of these events as more than half of marine species and nearly one fifth of terrestrial vertebrates, including dinosaurs, all perished.

New chapter added to the history of mankind

In the latter half of the 1970s the geologist Walter Alvarez, and his father Luis Alvarez, the Nobel-prize award-winning physicist, found an odd layer of iridium rich clay with a 30x concentration - deposited on the earth's crust. Iridium is among the most rare elements on Earth and the majority of it fell to the bottom as the planet formed. However, it's commonly found in meteorites as well as an Asteroid, which suggests that one of them may have brought it to the region. Based on its location in the rock, they discovered that it was dumped at a time of 65 million years. This coincided with the time when the dinosaurs, along with a variety of other species of animals and plants were thought to have vanished.

As cosmic iridium rains through the atmosphere before settling within the soil, the amount of iridium found in the sediment layer can be used as a kind as cosmic counter clock. At any time, the amount of iridium will be mixed with soil, and eventually become part of the soil layer which forms. A layer that takes twice as long to develop can vary, however the layer that takes two times as long to form will have more iridium. However, 30x the normal amount is not enough to be considered a coincidence. In addition, this rich deposit was discovered on more than 100 different sites all over the world, suggesting an event that is truly global.

The researchers' Alvarez and their colleagues found that the flood of iridium originated from the other side of the Earth. In order to deposit such huge quantities the impactor would need to be large enough that the material was blasted upwards along with other debris and then transported around the globe by strong winds. The object must to be a few kilometers in diameter. Evidence was found after impact-shocked quartz was impacted and crystals that were made from glass impact discovered within the layer of clay. These substances are produced by high-pressure shock waves and are discovered in impact craters as well as nuclear bomb locations. Very few terrestrial processes such as volcanic flows or explosions produce the power and temperature needed, which is why the possibility of an impact by an asteroid was suggested. This left a clear issue: where was the impact's crater located that destroyed the majority of living things on Earth?

After years of looking an enormous crater was found located in Chicxulub (a Mayan village, which is pronounced "Chik-shoo-loob" meaning "horns of devils") located in the Yucatan peninsula in Mexico. With an area of 180 km, and with a date of the age of 65 million It was of the correct dimension and age which led into the acceptance of the impact hypothesis as a reason for the extinction of dinosaurs.

Figure 9.1 The topography in radar images taken from Space Shuttle Endeavour. Space Shuttle Endeavour depicting the 180-kilometer Chicxulub crater located on the Yucatan Peninsula. Sinkholes, also known as cenotes (pronounced "sen-o-tees"), can be found all over across the Yucatan peninsula, however it is the cenotes that have an obvious connection in the Chicxulub crater. There are numerous sinkholes that are aligned in circular patterns that mark the rim of the crater that helped identify the crater's place.

The initial blast destroyed everything within the immediate area and triggered a massive flameball that heated the oceans and generated winds in more than 400 kilometres an hour! The shockwave destroyed forests across North as well as South America. Evidence of a massive tsunami can be seen present in Spain and Brazil and it is possible that the tsunami could be reaching New Zealand!

Then, the pain really got worse!

Figure 9.2 The final day of the world about 65 million years ago.

Credit: Don Davis, NASA

Despite the magnitude that the impact had initially caused it was the lasting effects that caused the mass extermination of the species. The pulverised debris that was ejected into the air would be swept around by air currents until it covered the Earth with a cloak of darkness. By blocking light and heat The temperature on the surface plummeted and remained that way for months.

In the absence of sunlight, plant life are unable to photosynthesis, and consequently will wilt and die. Animals that eat plants will quickly disappear, as would meat eaters when their prey had gone. As the dust fell, species of land plants and animals were gone.

Other devastations could have included huge wildfires that would have occurred when the hot material returned to Earth in the form of super-heating the atmosphere and causing a massive firestorm. A toxic substance blown into air could acidify the rain and impure the water, killing fish and creatures that eat shells.

In time, the planet was able to heal, but dinosaurs were gone along with many species of marine animals and plants; perhaps the majority of animals and plants were eliminated.

Where there is chaos, there's also the chance. The demise of one species allows others to succeed as well, for instance, humans have flourished in the last 500 million years. Our ancestors, apes, were our first ancestors and humans, we've created civilization that has built huge cities, and ultimately, explored beyond our own world and into the space. Could it be that the world is coming in the end? In the short time we live the possibility of "death through Asteroid" is not that high However, when you consider thousands or billions of years it's a sure thing. over 99percent of the species that ever existed have come to an end.

Cenotes Cenotes

Inside the Yucatan peninsula There is a massive complex of caves and tunnels known as The Cenotes (see figure 9.1). When viewed from the top, one can observe that they are scattered over hundreds of kilometers across all directions. But, when shown on a map they outline the outline of a circular trail through the forest - the rim of a huge crater. This is evident in part due to this alignment, but also by geophysical mapping, the pattern of the land beneath it is now revealed to be affected. The crater is recorded as having a age of 65 million years old at the time that what is known as"the Cretaceous (K) time period came to an end in addition to when that of the Paleogene (Pg) phase of the Tertiary era began. The meteorite therefore is referred to in the context of the Cretaceous-Tertiary (K/T) or the Cretaceous-Paleogene event of extinction.

The discovery of the 1980s revolutionized how we looked at the Earth. Prior to that the idea was that the changes in Earth's surface

Earth were all incredibly slow processes, however it was now clear that rapid and abrupt disasters had affected the Earth as well. Another, possibly more significant concern was the possibility that if this had occurred once, it might be repeated. Luckily, the largest asteroids are far and few from each other, however there are numerous smaller rocks in the vicinity that pose the threat of a serious threat to us.

TC3 2008

In in the wee hours on sixth of October, an asteroid hunter Richard Kowalski spotted something that could change the way we view the threat that asteroids pose to us. The night was going as planned but, when he stepped onto his screen, appeared an asteroid that was new. It was not atypical in itself however, he sat and watched its movements throughout the night, and noticed that its movements on the screen were increasing in speed, which indicated that it was extremely close.

These observations, as well as many other ones, were reported and then compiled by the very heart of asteroid detection, The Minor Planets Centre (MPC). Computers ran around and crunched the numbers however, they were not able to calculate an orbit. Scientists realized that the object was on a collision path with Earth. Earth and, more ominously the fact that it was scheduled to strike within 19 hours!

In accordance with a strict procedure developed to deal with such situations Following a strict protocol, the MPC immediately reached out to NASA's Asteroid Investigation team in California who gathered the data and verified the 100% likelihood of impact on the next day. This kind of event has never been observed except for testing and simulation.

Astronomical impacts could trigger an enormous and destructive explosion. In fact, it could even be mistaken for the nuclear bomb! NASA has informed of the White House and let them know that the asteroid was headed for them. The object was too close to avoid the strike. all that was left was to figure out the exact location it would strike.

NASA predicted an area of desert that was isolated in that region, Nubian Desert in the northern Sudan and at third to three in the daytime,, they were proven correct. The explosion created a huge explosion that was as bright as the Sun and even was captured by an orbiting satellite for weather. It is somewhat alarming that this object measured just 4 metres in size - significantly smaller than the one that was observed over Chelyabinsk.

Figure 9.3: The explosion that occurred in the Nubian Desert, thanks to the asteroid 2008TC3. A thin line of blue that runs through the middle of the Nile. In the northern part is Egypt and to the east lies it is the Red Sea.

The thing that was alarming about this particular event was the fact that the asteroid was too small to be observed until very late in the daytime. This means that defending against these objects is almost impossible. These rocks are not a threat to Earth and yet, only a few meters across, they possess the explosive force of a tiny nuclear bomb.

Encouragingly, we are getting better at detecting asteroids. Some have even come close in between Earth with the Moon (384,400 km). In the last few years there have been more than 100 that have come close to this distance with the majority of them less than 200m in size. While there is plenty of space between us and Moon that asteroids can pass through, when you consider the size of space overall the distance between us and the Moon is extremely small. Some come even closer.

Do asteroids exist in twos?

On February 15, the exact day on which the Chelyabinsk meteorite landed the same day, another body, 2012 DA14 was flying by the Earth. At a distance of just 28,000 km away from Earth's surface Many stories on the news claimed that the two objects were similar, even travelling across space in tandem. However, in reality they were totally unrelated in that they came from different directions. It was just a coincidence that they arrived at the same at the same time. The first object was spotted from the outer edge of the asteroid belt the other one followed an extremely evolved orbit close to the Earth.

The same size as the one that caused the Barringer crater. It was spotted from below the surface of Earth and travelled through many spacecrafts prior to moving towards the North. The asteroid was observed and tracked for over a year, and despite its proximity to Earth, scientists were aware that it wasn't a threat.

This rock threads is a link connects Low Earth Orbit (LEO) which is where is where the

International Space Station and numerous observation satellites are situated in the lower altitude, and the Geostationary Earth Orbit (GEO) where satellites for communications and weather are located.

Figure 9.4 The Lower Earth Orbits (LEOs) are those that lie between 200 and 2000 km, in which the bulk of spacecraft and satellites happens. Spacecraft operating in this region orbit Earth within 90 minutes. In this region, the International Space Station lies at 400 km. Its close proximity to Earth means that communication with Earth significantly easier than GEO orbits, however orbits here are constantly losing the altitude (decay) and have to be periodically boosted to higher altitude.

Credit: Apollo 17/ NASA

Figure 9.5 9. Geostationary Earth orbit is one in which it orbits just above and below the Equator along the same direction Earth's spin as well at the exact rate which allows it to maintain a constant location within the skies. In these orbits, around 36,000 km above the earth's surface the satellites for weather and communications are situated.

Note: Geostationary orbits are specific instances of Geosynchronous Orbits. They are in the exact same elevation and are orbiting in the same time frame, however they aren't directly above the equator; rather, they change position constantly relative to the ground beneath.

Figure 9.6 The close to the surface of 2012 DA 14 through the plane of the ecliptic. All dates were in GMT on 15th (and 16th) February 2013.

In 2013, when it was on the 31st day of May when the Asteroid 1998 QE2 flew gently over the Earth. The asteroid was not close to 5.8 million kilometers (about 15 distances between Earth and Moon) it was not of much significance to astronomers or researchers worried about the dangers of asteroids. However at 2.7 km it was a massive one that was too large to be ignored completely. because of that it proved the ease of objects to be overlooked without being directly examined. Like the name suggests the discovery was made fifteen years ago, but it wasn't until the moment it came close that radar imaging uncover something completely unusual: it has its moon! This, on its own isn't a huge surprise as around 16% of objects in the Near-Earth space that are greater than 200 metres in diameter belong to the triple or double system. Larger objects, which have more gravitation will be more likely to grasp the object they travel by. In this case the moon was massive 600 meters across. However, if it wasn't for the astronomers who were looking directly at the larger moon and the smaller one, it could be a complete mystery. Although we saw this one a long distance away, other aren't so obvious. On the 17th of June 2002 an asteroid measuring 100m from end to end traveled just under a third of the distance toward the Moon. Surprisingly, it was discovered just three days after it had passed!

How many of them float over us, and remain unseen?

Things that didn't occur at Chelyabinsk

What was the most shocking aspect of the incident was what didn't happen There was no fear about this being an attack on nuclear weapons. The year was 1981. Gene Shoemaker warned that unintentional meteorite impacts could be misinterpreted as an attack, and could trigger an outbreak of nuclear conflict. Certainly, during the height of the Cold War, this was an actual fear and is still a possibility if an incident to occur amid the conflicting territories that are occupied by nuclear-armed enemies such as India as well as Pakistan.

Maybe at Chelyabinsk There is even another reason to be wary of attack. In the vicinity there are a number of Russian defense facilities and some of them are also building nuclear weapons. In 1960 in 1960, the U.S. U-2 flight of Gary Powers was shot down by the Soviets close to the region. The truth is, neither the public or the military believed that the region was under attack and we are able to be grateful.

A death from the sky: What do you think are the odds?

A few hundred miles above, asteroids roaming around will soon be able to join the Earth. Anything that is large enough will hurt... quite a bit. We do not know when or where they'll land the ground, but we can determine the chance. Large rocks can cause a lot of damage, but are very uncommon. Small rocks are abundant, and they are more likely to be hit. By sorting them according to dimension, one can determine the probability of being struck by something large. The majority of them are less than 10s of meters and are able to burn in the air, but small pieces could hit the earth. As we've seen although small rocks may be far more dangerous than was initially thought however, they aren't a danger to the Earth in general. It is only when we get into rocks that are more than a kilometre across can we contemplate end-of-the world scenarios.

Figure 9.9 Probabilities of Impact. This diagram illustrates the probabilities for an item of particular size striking the Earth and the destruction it could cause. The frequency of the impact is shown on the vertical axis together with the dimensions (and the force) of each object plotted on the horizontal. Such objects as TC3 2008 are fairly frequent, hitting us each year. A similar object that destroyed dinosaurs is seen about each 100 million years. A 40-50 meters object similar to that found at Tunguska is only once in a millennium. In their own terms the explosives' values are nothing, and to place them in perspective the nuclear weapon used on Hiroshima in 1945 was an explosive result that was "only" 16 kilograms of TNT. This is roughly what the annual 4 meter space-based impacts! In addition, the biggest bomb that ever went off in history was Tsar Bomb, tested by the USSR in the year 1961. This yielded around 50 megatons TNT in contrast, the Tunguska incident was possibly only 30 megatons! It is believed that the threshold of a global disaster is somewhere within the range of 1-2 kilometers, which makes an event that is only once in a million years.

Recent surveys suggest that there could be a thousand pieces per km or more in transit or crossing our orbit. It is believed that the 10,000th NEO was discovered recently however, 90% of them are not more than a kilometre in size and even one or two feet The largest, 1036 Ganymed 41 km long.

Note: "Ganymed", is the German spelling of "Ganymede" and is the name Jupiter's biggest moon.

For millions of years, they float in space, focusing on their work however, occasionally, one's path might be blocked. It's true that "death caused by an the asteroid" is not a primary priority for the majority of people, possibly with good reasons. However, in order to put things in perspective, here are some interesting facts we can look at.

The interval between impacts of major magnitude is approximately 1 million years, far more than the time of our civilisation. It is interesting that in a lifespan of 65 years that the odds of being killed by an asteroid strike are thought to be 1 in 20 000 . That's exactly the same as being killed in the event of a plane crash! Of course, nobody was ever killed in the event of an object, however the event that one does occur with a devastating impact, it could take hundreds of millions or thousands of life, significantly increasing the likelihood of death in one. In addition, no 1 km or larger object (that we are aware of) is expected to strike us in the next century or anytime soon.

However, there are numerous NEOs in the universe which are smaller than 1km in size, with the amount of objects growing with their size decreasing. There are about 15,000 NEOs approximately 140 meters in size (about one-and-a-half times the dimension of the football field). Over a million are approximately 30 meters in size approximately the dimension that would cause serious damage to areas of habitation.

The issue lies in the fact that the less the items are the more difficult they are to locate. A little over 30 percent of the 140-metre objects have been found however, less than 1 percent of the bodies measuring 30 metres are identified, and if any of them was to arrive, there will be only a few days or even no warning.

At present at present, we are at a point where Earth isn't in danger of imminent catastrophe. There are no massive objects scheduled to collide in the next century and we are able to trace objects to 10 metres in size. What about objects that are smaller than that, like the one that burst into flames over Chelyabinsk? We have next chances of securing ourselves from such. The possibility of one could arise next day, or in the next decade as well, and until we are able to find, catalog and monitor all potential dangersome objects within our Solar System, it is an opportunity that we always be at risk.

...And there's a third aspect that can make asteroids difficult to find and is why nobody has noticed the object speeding toward Chelyabinsk because it was straight into the Sun. There are numerous telescopes that search for stars in the sky however, they are only operational in the evening.